COLOUR CHEMISTRY

RSC Paperbacks

RSC Paperbacks are a series of inexpensive texts suitable for teachers and students and give a clear, readable introduction to selected topics in chemistry. They should also appeal to the general chemist. For further information on all available titles contact:

Sales and Customer Care Department, Royal Society of Chemistry,
Thomas Graham House, Science Park, Milton Road, Cambridge CB4 0WF, UK
Telephone: +44 (0)1223 432360; Fax: +44 (0)1223 423429; E-mail: sales@rsc.org

Recent Titles Available

The Chemistry of Fragrances
compiled by David Pybus and Charles Sell
Polymers and the Environment
by Gerald Scott
Brewing
by Ian S. Hornsey
The Chemistry of Fireworks
by Michael S. Russell
Water (Second Edition): A Matrix of Life
by Felix Franks
The Science of Chocolate
by Stephen T. Beckett
The Science of Sugar Confectionery
by W. P. Edwards
Colour Chemistry
by R. M. Christie

Future titles may be obtained immediately on publication by placing a standing order for RSC Paperbacks. Information on this is available from the address above.

RSC Paperbacks

COLOUR CHEMISTRY

R. M. CHRISTIE

Heriot-Watt University, Scottish Borders Campus, Galashiels, UK

RS•C

ROYAL SOCIETY OF CHEMISTRY

ISBN 0-85404-573-2

A catalogue record for this book is available from the British Library

Published by The Royal Society of Chemistry,
Thomas Graham House, Science Park, Milton Road,
Cambridge CB4 0WF, UK

For further information see our web site at www.rsc.org

Typeset in Great Britain by Vision Typesetting, Manchester
Printed by Bookcraft Ltd, UK

Preface

This book provides an insight into the chemistry of colour. It is aimed primarily at students or graduates who have a knowledge of the principles of chemistry, to provide an illustration of how these principles are applied in producing the range of colours which are all around us. In addition, it is anticipated that readers who are specialists in colour science, or have some involvement in an industrial or academic environment with the diverse range of coloured materials, will benefit from the overview of the subject provided.

The scene is set with an opening chapter giving a historical perspective on how our understanding of colour science has evolved. The second chapter provides a general introduction to the physical, chemical, and to a certain extent, biological principles which allow us to perceive colours. Chapters 3 to 6 encompass the essential principles of the structural and synthetic chemistry associated with the various chemical classes of dyes and pigments. Chapters 7 to 10 deal with the application of dyes and pigments, and in particular the chemical principles underlying their technical performance not only in traditional applications such as textiles, coatings and plastics but also in an expanding range of high technology or functional applications. The final chapter provides a selective account of some of the important environmental issues associated with the manufacture and use of colour.

The book concludes with a bibliography. The author expresses his gratitude to the authors of these texts for providing sources both of information and of inspiration.

R. M. Christie

Contents

Abbreviations Used

A	Acceptor
C.I.	Colour Index or Configuration interaction
CuPc	Copper phthalocyanine
D	Donor
D2T2	Dye diffusion thermal transfer
DCB	Dichlorobenzidine
DPP	Diketopyrrolopyrrole
FBAs	Fluorescent brightening agents
FePc	Iron phthalocyanine
HMO	Hückel molecular orbital
PCB	Polychlorobiphenyl
PDT	photodynamic therapy
PET	polyethylene terephthalate
PPP	Pariser–Pople–Parr
PPV	Poly(p-phenylenevinylene)
SCF	Self-consistent field
SDC	Society of Dyers and Colourists
UV	Ultraviolet
VSIP	Valence state ionisation potential

Chapter 1

Colour: A Brief Historical Perspective

We only have to open our eyes and look around to observe how important a part colour plays in our everyday lives. Colour influences our
moods and emotions and generally enhances the way in which we enjoy
our surroundings. Our experience of colour emanates from a rich diversity of sources, both natural and synthetic. Natural colours are all around
us, in the earth, the sky, the sea, animals and birds and in the vegetation,
for example in the trees, leaves, grass and flowers. Colour is an important
aspect in our enjoyment of the food we eat. In fact, we frequently judge the
quality of meat products, fruit and vegetables by the richness of their
colour. In addition, there is a myriad of examples of synthetic colours,
products of the chemical manufacturing industry, which we tend to take
so much for granted these days. These colours commonly serve a purely
decorative or aesthetic purpose, but in some cases specific colours may be
used to convey vital information, for example in traffic lights and colour-
coded electrical cables. Synthetic colours are used in the clothes we wear,
in paints, plastic articles, in a wide range of multicoloured printed
material such as posters, magazines and newspapers, in photographs,
cosmetics, ceramics, and on television and film. Colour is introduced into
these materials using substances known as dyes and pigments. The essential difference between these two types of colorants is that dyes are soluble
coloured compounds which are applied mainly to textile materials from
solution in water, whereas pigments are insoluble compounds incorporated by a dispersion process into products such as paints, printing inks
and plastics. The reader is directed to Chapter 2 of this book for a more
detailed discussion of the distinction between dyes and pigments as
colouring materials.

People have made use of colour since prehistoric times, for example in
decorating their bodies, in colouring the furs and skins that they wore and
in the paintings which adorned their cave dwellings. Of course, in those

1

days the colours that were used were derived from natural resources. The dyes used to colour clothing were commonly extracted either from vegetable sources, including plants, trees, roots, seeds, nuts, fruit skins, berries and lichens, or from animal sources such as crushed insects and molluscs. The pigments for the paints were obtained from coloured minerals, such as ochre and haematite, which were dug from the earth, ground to a fine powder and mixed into a crude binder.

Synthetic colorants may also be described as having an ancient history, although this statement applies only to a range of pigments produced from basic applications of inorganic chemistry. These very early synthetic inorganic pigments have been manufactured and used in paints for many thousands of years. The ancient Egyptians were probably responsible for the development of the earliest synthetic pigments. The most notable products were Alexandra blue, a ground glass coloured with a copper ore, and Egyptian Blue, a mixed silicate of copper and calcium which has been identified in murals dating from around 1000 BC. Perhaps the oldest synthetic colorant still used extensively today is Prussian Blue, the structure of which has been established as iron(III) hexacyanoferrate(II). The manufacture of this blue inorganic pigment is much less ancient, dating originally from the middle of the 17th century, although this product pre-dates the origin of synthetic organic dyes and pigments by more than a century.

Synthetic textile dyes are exclusively organic compounds and, in relative historical terms, their origin is much more recent. Textile materials were coloured exclusively with the use of natural dyes until the mid-19th century. Since most of nature's dyes are rather unstable, the dyeings produced in the very early days tended to be quite fugitive, for example to washing and light. Over the centuries, however, complex dyeing procedures using a selected range of natural dyes were developed which were capable of giving reasonably quality dyeing on textile fabrics. Since natural dyes generally have little direct affinity for textile materials, they were usually applied together with compounds known as *mordants*, which were effectively 'fixing-agents'. Metal salts, for example of iron, tin, chromium, copper or aluminium, were the most commonly used mordants, and these functioned by forming complexes of the dyes within the fibre. These complexes were insoluble and hence more resistant to washing processes. As a result, these agents not only improved the fastness properties of the dyeing, but also in many instances were essential to develop the intensity and brightness of the colours produced by the natural dyes. Some natural organic materials such as tannic and tartaric acids could also be used as mordants. Among the most important of the natural dyes the use of which has been sustained over the centuries, is indigo **1a**, a blue dye obtained from certain plants, for example from *Indigofera tinctoria* found in India, and from woad, a plant extract. A related product is Tyrian purple, whose principal constituent was 6,6′-dibromoindigo **1b**. This was for many years a fashionable purple dye which was extracted from the glands of *Murex brandaris*, a shellfish found on the Mediterranean and Atlantic coasts. The most important of the natural red dyes was madder, a wood extract, the main constituent of which was alizarin, 1,2-dihydroxyanthraquinone (**2**). Alizarin provides a good example of the use of the mordanting process, since it readily forms metal complexes within fibres, notably with aluminium, which show more intense colours and an enhanced set of fastness properties.

1

1a R = H; **1b** R = Br

2

It may be argued that the first synthetic dye was picric acid **3**, which was first prepared in the laboratory in 1771 by treating indigo with nitric acid. Much later, a more efficient synthetic route to picric acid from phenol as the starting material was developed. Picric acid was found to dye silk a bright greenish-yellow colour but it did not attain any real significance as a practical dye mainly because the dyeings obtained were of poor quality, especially in terms of lightfastness. However, it did find limited use at the time to shade indigo dyeings to give bright greens.

3

The foundation of the synthetic dye industry is universally attributed to William Henry Perkin on account of his discovery in 1856 of a purple dye which he originally gave the name Aniline Purple, but which was later to become known as Mauveine. Perkin was a young enthusiastic British organic chemist who was carrying out research aimed not initially at synthetic dyes but rather at developing a synthetic route to quinine, the antimalarial drug. His objective in one particular set of experiments was to prepare synthetic quinine from the oxidation of allyltoluidine, but his attempts to this end proved singularly unsuccessful, and, with hindsight, this is not too surprising in view of our current knowledge of the complex heteroalicyclic structure of quinine. As an extension of this research, he turned his attention to the reaction of the simplest aromatic amine, aniline, with the oxidising agent potassium dichromate. This reaction gave a black product which to many chemists might have seemed rather unpromising, but from which Perkin discovered that a low yield of a purple dye could be extracted with solvents. An evaluation of the dye in a silk dyeworks in Perth, Scotland, established that it could be used to dye silk a rich purple colour and give reasonable fastness properties. Perkin showed remarkable foresight in recognising the potential of his discovery. He took out a patent on the product and had the boldness to instigate the development of a large-scale manufacturing process. The dye was launched on the market in 1857. Since the manufacture required the development of large-scale industrial procedures for the manufacture of aniline from benzene *via* reduction of nitrobenzene, the real significance of Perkin's venture was as the origin of the organic chemicals industry, which has evolved from this humble beginning to become a dominant feature of the industrial base of many of the world's developed countries. For many years, the structure of Mauveine was reported erroneously as **4**. It has been demonstrated only relatively recently from an analytical investigation of an original sample that the dye is a mixture and that the structures of the principal constituents are in fact compounds **5** and **6**. The presence of the methyl groups, which are an essential feature of the product, demonstrate how fortuitous it was that Perkin's crude aniline contained significant quantities of the toluidines. Compound **5**, the major component of the dye, is derived from two molecules of aniline, one of *p*-toluidine and one of *o*-toluidine, while compound **6** is formed from one molecule of aniline, one of *p*-toluidine and two molecules of *o*-toluidine.

During the several years following the discovery of Mauveine, research activity in dye chemistry intensified, especially in Britain, Germany, and France. For the most part, chemists concentrated on aniline as the starting material, adopting a largely empirical approach to its conversion into coloured compounds, and this resulted in the discovery of several other synthetic textile dyes with commercial potential. In fact the term 'Aniline Dyes' was for many decades synonymous with synthetic dyes. Most notable among the initial discoveries were the triphenylmethine dyes, the first important commercial example of which was Magenta, introduced in 1859. Magenta was first prepared by the oxidation of crude aniline (containing variable quantities of the toluidines) with tin(IV) chloride. The dye contains two principal constituents, rosaniline **7** and

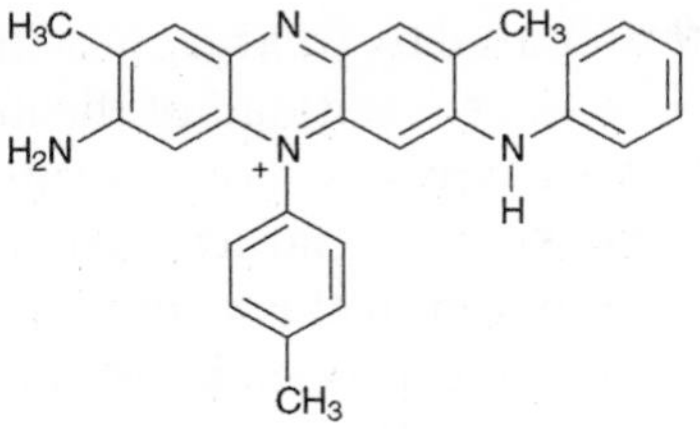

4

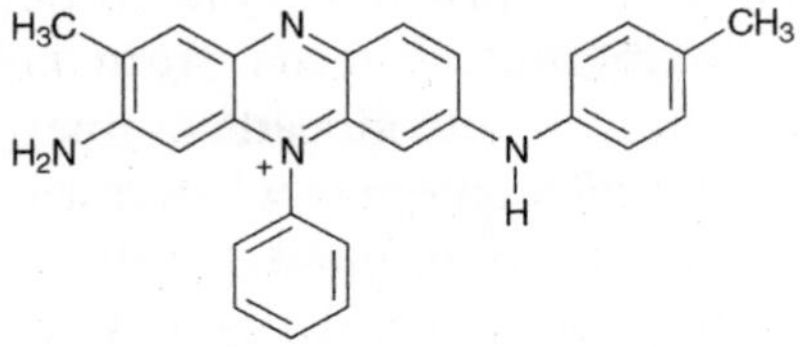

5

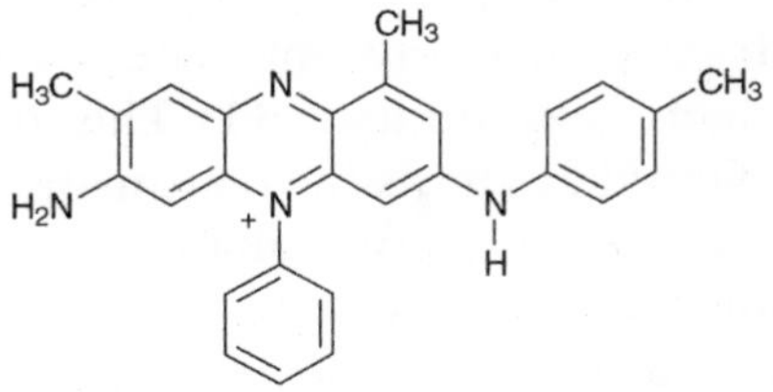

6

homorosaniline **8**, the central carbon atom being derived from the methyl group of *p*-toluidine. A structurally related dye, rosolic acid had been prepared in the laboratory in 1834 by the oxidation of crude phenol, and therefore may be considered as one of the earliest synthetic dyes, although its commercial manufacture was not attempted until the 1860s. Structure **9** has been suggested for rosolic acid, although it seems likely that other components were present. A wide range of new triphenylmethine dyes soon emerged and these proved quite quickly to be superior in properties and were more economic compared with Mauveine, the production of which ceased after about ten years. All of these dyes, including Mauveine, may be considered as examples of the arylcarbonium ion chemical class of dyes, many of which are still of current importance (see Chapter 6).

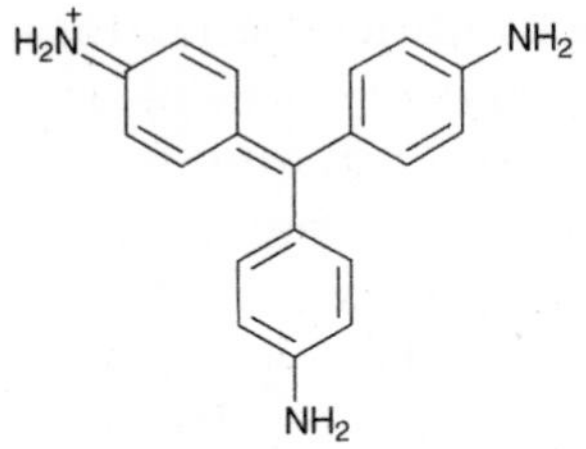

7

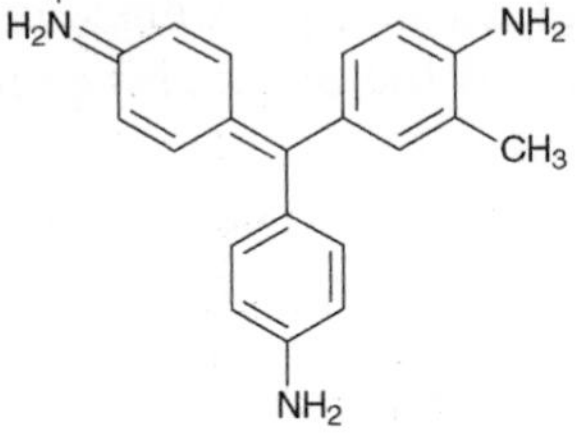

8

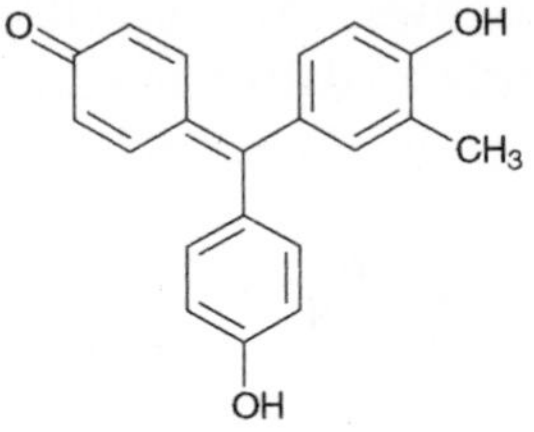

9

Undoubtedly the most significant discovery in colour chemistry in the 'post-Mauveine' period was due to the work of Peter Griess, which provided the foundation for the development of the chemistry of azo dyes and pigments. In 1858, Griess demonstrated that the treatment of an aromatic amine with nitrous acid gave rise to an unstable salt (a diazonium salt) which could be used to prepare highly coloured compounds.

The earliest azo dyes were prepared by treatment of aromatic amines with a half-equivalent of nitrous acid, so that half of the amine was diazotised and the remainder acted as the coupling component in the formation of the azo compound. The first commercial azo dye was 4-aminoazobenzene (**10**), Aniline Yellow, prepared in this way from aniline, although it proved to have quite poor dyeing properties. A much more successful commercial product was Bismarck Brown, (originally named Manchester Brown) – actually a mixture of compounds, the principal constituent of which is compound **11**. This dye was obtained directly from *m*-phenylenediamine as the starting material and was introduced commercially in 1861. The true value of azo dyes eventually emerged when it was demonstrated that different diazo and coupling components could be used, thus extending dramatically the range of coloured compounds which could be prepared. The first commercial azo dye of this type was Chrysoidine which was derived from reaction of diazotised aniline with *m*-phenylenediamine and was introduced to the market in 1876. This was followed soon after by a series of orange dyes (Orange I, II, III and IV) which were prepared by reacting diazotised sulfanilic acid (4-aminobenzene-1-sulfonic acid) respectively with 1-naphthol, 2-naphthol, *N,N*-dimethylaniline and diphenylamine. In 1879, Biebrich Scarlet, **12**, the first commercial disazo dye to be prepared from separate diazo and coupling components, was introduced. History has demonstrated that azo dyes were to emerge as by far the most important chemical class of dyes and pigments, dominating most applications (see Chapter 3). It was becoming apparent that the synthetic textile dyes which were being developed were cheaper, easier to apply and were capable of providing better colour and technical performance than the range of natural dyes applied by traditional methods. As a consequence, within 50 years of Perkin's initial discovery, around 90% of textile dyes were synthetic rather than natural and azo dyes had emerged as the dominant chemical type.

Towards the end of the 19th century, a range of organic pigments was also being developed commercially, particularly for paint applications. These were found to provide brighter, more intense colours than the inorganic pigments which had been in use for many years. Initially, most of these organic pigments were obtained from established water-soluble textile dyes. Anionic dyes were rendered insoluble by precipitation onto inert colourless inorganic substrates such as alumina and barium sulfate while, alternatively, basic dyes were treated with tannin or antimony potassium tartrate to give insoluble pigments. Such products were commonly referred to as 'lakes'. Their introduction was followed soon after by the development of a group of yellow and red azo pigments, such as the Hansa Yellows and β-naphthol reds, which did not contain substituents capable of salt formation. Many of these products are still of considerable importance today, and are referred to commonly as the classical azo pigments (see Chapter 9).

It is of interest and in a sense quite remarkable that, at the time of Perkin's discovery of Mauveine, chemists had very little understanding of the principles of organic chemistry. As an example, even the structure of benzene, the simplest aromatic compound, was an unknown quantity. Kekulé's proposal concerning the cyclic structure of benzene in 1865 without doubt made one of the most significant contributions to the development of organic chemistry. It is certain that the commercial developments in synthetic colour chemistry which took place from that time onwards owed much to the coming of age of organic chemistry as a science. For example, the structures of some of the more important natural dyes, including indigo (**1a**) and alizarin (**2**), were deduced. In this period well before the advent of the modern range of instrumental analytical techniques which are now used routinely for structural analysis, these deductions usually arose from a painstaking investigation of the chemistry of the dyes, commonly involving a planned series of degradation experiments from which identifiable products could be isolated. Following the elucidation of the chemical structures of these natural dyes, a considerable amount of research effort was initiated which was devoted to devising efficient synthetic routes to these products. The synthetic routes which were developed for the manufacture of these dyes proved very quickly to be significantly more cost-effective than the traditional methods which involved extracting the dyes from natural sources and in addition gave the products more consistently and with better purity. At the same time, by exploring the chemistry of these natural dye systems, chemists were discovering a wide range of structurally related dyes which could be produced synthetically and which had excellent colour properties and technical performance. As a consequence, the field of carbonyl dye chemistry, and the anthraquinones in particular, had opened up and this group of dyes remains for most textile applications the second most important chemical class, after the azo dyes, in use today (see Chapter 4).

   In the first half of the 20th century new chemical classes of organic dyes and pigments continued to be discovered. Probably the most significant discovery was that of the phthalocyanines, which have become established as the most important group of blue and green organic pigments. As with virtually every other new type of chromophore developed over the years, the discovery of the phthalocyanines was fortuitous. In 1928, chemists at Scottish Dyes, Grangemouth observed the presence of a blue impurity in certain batches of phthalimide produced from the reaction of phthalic anhydride with ammonia. They were able to isolate the blue impurity and subsequently its structure was established as iron phthalocyanine. The source of the iron proved to be the reactor vessel wall, which had become exposed to the reactants as a result of a damaged glass lining. Following this discovery, the chemistry of formation of phthalocyanines and their chemical structure and properties was investigated extensively by Linstead of Imperial College, London. Copper phthalocyanine (**13**) emerged as by far the most important product, a blue pigment which is capable of providing a brilliant intense blue colour and excellent technical performance, yet which at the same time can be manufactured at low cost in high yield from commodity starting materials (see Chapter 5). The discovery of this unique product set new standards for subsequent developments in dye and pigment chemistry.

13

   As time progressed, the strategies adopted in dye and pigment research evolved from those used in the early approach which were based largely on empiricism and involved the synthesis and evaluation of large numbers of products, to a more structured approach involving more fundamental studies of chemical principles. For example, attention turned to the reaction mechanisms involved in the synthesis of dyes and pigments and to the interactions between dye molecules and textile fibres. Probably the most notable advance in textile dyeing which has taken place during the 20th century and which, it may be argued, emerged from such investigations is the process of reactive dyeing. Reactive dyes contain functional groups which, after application of the dyes to certain fibres, can be linked covalently to the polymer molecules that make up the fibres, and this gives rise to dyeings with superior washfastness compared with the more traditional dyeing processes. Dyes which contain the 1,3,5-triazinyl group, discovered by ICI in 1954, were the first successful group of fibre-reactive dyes. The introduction of these products to the market as Procion dyes by ICI in 1956, initially for application to cellulosic fibres such as cotton, proved to be a rapid commercial success. The chemistry involved when Procion dyes react with the hydroxyl groups present on cellulosic fibres under alkaline conditions involves the aromatic nucleophilic substitution process outlined in Scheme 1.1, in which the cellulosate anion is the effective nucleophile. Reactive dyes have developed into one of the most important application classes of dyes for cellulosic fibres, and their use has been extended to a certain extent to other types of fibres, notably wool and nylon (see Chapter 8).

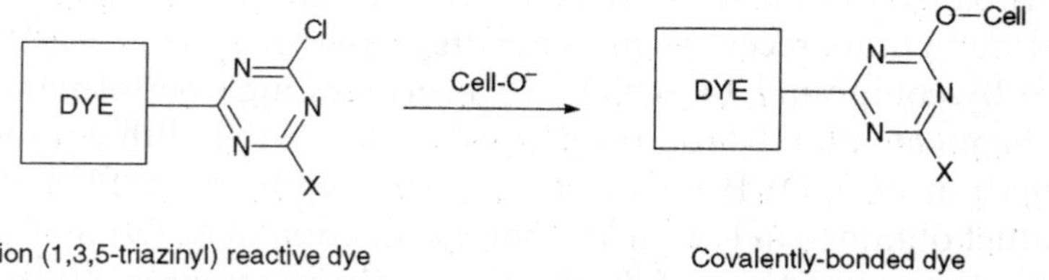

Procion (1,3,5-triazinyl) reactive dye   Covalently-bonded dye

Scheme 1.1   *The reaction of Procion dyes with cellulosic fibres*

In the latter part of the 20th century, new types of dyes and pigments for the traditional applications of textiles, leather, plastics, paints and printing inks continued to be developed and introduced commercially but at a declining rate. It is clear that the colour manufacturing industries considered that a mature range of products existed for these conventional applications and elected to transfer research emphasis towards process and product development, and the consolidation of existing product ranges. At the same time, during this period, research effort in organic colour chemistry developed in new directions, sustained by the opportunities presented by the emergence of a range of novel applications demanding new types of colorants. These colorants have commonly been termed 'functional dyes' because the applications require the dyes to perform certain functions beyond simply providing colour. The applications of functional dyes include some of the more recently developed reprographic techniques, such as electrophotography and ink-jet printing, a wide range of electronic applications including optical data storage, liquid crystal displays, lasers and solar energy conversion, and a range of medical uses (see Chapter 10). For these new applications, and also for traditional uses, concepts of molecular design of new dyes for specific properties have become increasingly important. Of particular significance in this respect is the fact that quantum mechanical methods have become accessible as a routine tool for the calculation of a range of properties of dyes, facilitated by the rapid advances in computing technology. The Pariser–Pople–Parr (PPP) molecular orbital method has proved of particular value for this purpose, although a range of more sophisticated methods are becoming increasingly accessible. From a knowledge of the molecular structure of a dye, these methods may be used with a reasonable degree of confidence to predict by calculation its colour properties, including the hue of the dye from its calculated absorption maximum, and its intensity as indicated by its molar extinction coefficient. The availability of these methods allows the potential properties of large numbers of molecules to be screened as an aid to the selection of synthetic target molecules (see Chapter 2).

Probably the most important new chromophoric system to be discovered in more recent years is the diketopyrrolopyrroles (DPP), exemplified by compound **14**, which are a series of high performance brilliant red pigments that exhibit properties similar to the phthalocyanines. The formation of a DPP molecule was first reported in 1974 as a minor product obtained in low yield from the reaction of benzonitrile with ethyl bromoacetate and zinc. A fascinating study by research chemists at Ciba Geigy into the mechanistic pathways involved in the formation of the molecules led to the development of an efficient 'one-pot' synthetic procedure for the manufacture of DPP pigments from readily available starting materials (see Chapter 4). The development of DPP pigments has emerged as an outstanding example of the way in which an application of the fundamental principles of synthetic and mechanistic organic chemistry can lead to an important commercial outcome. There are other

14

important lessons for colour chemists in this discovery. It demonstrates that, well over a century after Perkin's discovery of Mauveine, there still remains scope for the development of new improved colorants for traditional colour applications. In addition, while powerful sophisticated molecular modelling methods are now available to assist in the design of new coloured molecules, the perception to follow-up and exploit the fortuitous discovery of a coloured compound, perhaps as a trace impurity in a reaction, will remain a vital complementary element in the search for new dyes and pigments.

The increasing public sensitivity towards the environment has in recent years had a major impact on the chemical industry. There is no doubt that one of the most important challenges to chemists currently is the requirement to satisfy the increasingly stringent toxicological and environmental constraints placed on industry as a consequence not only of government legislation but also of public perception. Since the products of the colour industry are designed to enhance our living environment, it may be argued that this industry has a special responsibility to ensure that its products and processes do not have an adverse impact on the environment in its wider sense (see Chapter 11).

Chapter 2

The Physical and Chemical Basis of Colour

It has been said that the presence of colour requires three things: a source of illumination, an object to interact with the light which comes from this source and a human eye to observe the effect which results. In the absence of any one of these, it may be argued that colour does not exist. A treatment of the basic principles underlying the origin of colour thus requires a consideration of each of these three aspects, which brings together concepts arising from three natural science disciplines, chemistry, physics and biology. Although the principal aim of this textbook is to deal with the chemistry of dyes and pigments, a complete appreciation of the science of colour cannot be achieved without some knowledge of the fundamental principles of the physical and biological processes which ultimately give rise to our ability to observe colours. This chapter therefore presents an introduction to the physics of visible light and the way it interacts with materials, together with a brief description of the physiology of the eye and how it responds to stimulation by light, thus giving rise to the sensation of colour. In addition, the chapter contains a discussion of some of the fundamental chemical principles associated with coloured compounds, including a description of how dyes and pigments may be classified, followed by an overview of the ways in which the chemical structure of a molecule influences its colour properties. This section places special emphasis on azo colorants because of their particular importance in the colour industry. These topics are presented as a prelude to the more detailed discussion of the chemistry of dyes and pigments contained in later chapters.

VISIBLE LIGHT

Visible light refers to the region of the electromagnetic spectrum to which our eyes are sensitive and corresponds to radiation within the very narrow wavelength range 360–780 nm. Since the sensitivity of the eye to radiation is very low at each of these extremes, in practice the visual spectrum is commonly taken as 380 to 720 nm. Beyond the extremes of this range are the ultraviolet (UV) region of the spectrum (below 360 nm) and the infrared (IR) region (above 780 nm).

Normal white light contains this entire wavelength range, although not necessarily in equal intensities. There are numerous sources of white light, some natural and some artificial in origin. The most familiar natural illumination is daylight, emanating from the sun. The visible light from the sun not only allows us to see objects, but it is in fact essential for life since it is the source of energy responsible for photosynthesis, the vital process which allows plants to grow and thus provides us with an essential food source. Normal daylight encompasses the complete visible wavelength range although its exact composition is extremely variable and dependent on a variety of factors such as the geographical location, the prevailing weather conditions, the time of day and the season. Artificial illuminants, such as the tungsten lamps and fluorescent lights that are used for interior lighting, are also sources nominally of white light, although the composition of the light from these sources varies markedly depending on the type of lamp in question. For example, tungsten lights appear yellowish as the light they emit is deficient in the blue region of the spectrum. Colours do appear different under different illumination sources, although when the human visual system assesses colours it is capable of making some allowance for the nature of the light source, for example by compensating for some of the deficiencies of artificial light.

The splitting of white light into its various component colours is a very familiar phenomenon. It may be achieved in the laboratory, for example, by passing a beam of white light through a glass prism, or naturally, as in a rainbow where the colours are produced by the interaction of sunlight with the raindrops. The visible spectrum is made up of specific wavelength regions that are recognised by the eye in terms of their characteristic colours. The approximate wavelength ranges of light corresponding to these observed colours are given in Table 2.1. Fundamental to the specification of colours is an understanding of the laws of colour mixing, the processes by which two or more colours are combined to 'synthesise' new colours. There are two fundamentally different ways in which this may be achieved: *additive* and *subtractive* colour mixing. Additive colour mixing, as the name implies, refers to the mixing of

Table 2.1 *Complementary colour relationships*

Wavelength range (nm)	Colour	Complementary colour
400–435	Violet	Greenish-yellow
435–480	Blue	Yellow
480–490	Greenish-blue	Orange
490–500	Bluish-green	Red
500–560	Green	Purple
560–580	Yellowish-green	Violet
580–595	Yellow	Blue
595–605	Orange	Greenish-blue
605–750	Red	Bluish-green

coloured lights, so that the source of illumination is observed directly by the eye. Subtractive colour mixing is involved when colours are observed as a result of reflection from or transmission through an object after it interacts with incident white light. The colours *red, green* and *blue* are referred to as the *additive primary colours*. Their particular significance is that they are colours that cannot be obtained by the mixing of lights of other colours, but they may be combined in appropriate proportions to produce the other colours. As illustrated in Figure 2.1(a), additive mixing of red and blue produces magenta, blue and green gives cyan, while combining red and green additively gives yellow. When all three primaries are mixed in this way, white light is created since the entire visible spectrum is present. A common situation in which additive colour mixing is encountered is in colour television, which uses separate red, green and blue emitting phosphors in the case of traditional cathode-ray tube technology, or using appropriately coloured microfilters in the case of flat-screen displays (see Chapter 10). When an object absorbs light of a given colour corresponding to its particular wavelength range, it is the

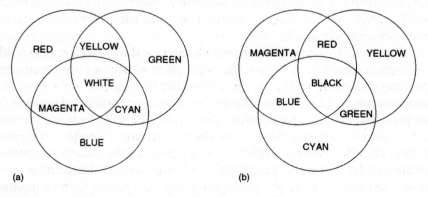

(a)   (b)

Figure 2.1   *(a) Additive colour mixing; (b) subtractive colour mixing*

complementary colour that is observed. The complementary colour corresponds to the remaining wavelengths of incident light, which are either transmitted or reflected, depending on whether the object is transparent or opaque, and are then detected by the eye. These complementary colour relationships are also given in Table 2.1. For example, an object that absorbs blue light (*i.e.* in the range 435–480 nm) will appear yellow, because the red and green components are reflected or transmitted. This forms the basis of subtractive colour mixing. This type of colour mixing, which is involved when dyes and pigments are mixed, is the more familiar of the two processes. The subtractive primary colours are *yellow*, *magenta* and *cyan*. These are the colours, for example, of the three printing inks used commonly to produce the vast quantities of multicolour printed material which we encounter in our daily lives, such as in magazines, posters, newspapers, *etc*. The principles of subtractive colour mixing are illustrated in Figure 2.1(b).

The colours described in Table 2.1 that are observed as a result of this selective light absorption process are referred to as *chromatic*. If all wavelengths of light are reflected from an object, it appears to the eye as white. If no light is reflected, we recognise it as black. If the object absorbs a constant fraction of the incident light throughout the visible region, it appears grey. White, black and grey are therefore referred to as *achromatic* since in those cases there is no selective absorption of light involved.

THE EYE

The sensation of colour that we experience arises from the interpretation by the brain of the signals that it receives *via* the optic nerve from the eye in response to stimulation by light. This section contains a brief description of the components of the eye and an outline of how each of these contributes to the mechanism by which we observe colours. Figure 2.2 shows a cross-section diagram of the eye, indicating some of the more important components.

The eye is enclosed in a white casing known as the *sclera*, or colloquially as the 'white of the eye'. The retina is the photosensitive component and is located at the rear of the eye. It is here that the image is formed by the focusing system. Light enters the eye through the *cornea*, a transparent section of the sclera, which is kept moist and free from dust by the tear ducts and by blinking of the eyelids. The light passes through a transparent flexible lens, the shape of which is determined by muscular control, and which acts to form an inverted image on the retina. The light control mechanism involves the *iris*, an annular shaped opaque layer, the inner diameter of which is controlled by the contraction and expansion of a set

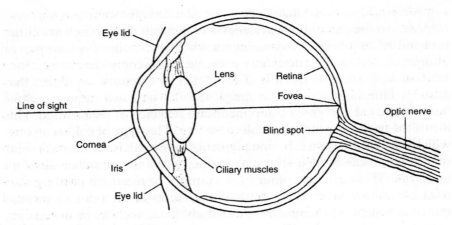

Figure 2.2   *Components of the eye*

of circular and radial muscles. The aperture formed by the iris is termed the *pupil*. Light passes into the eye through the pupil, which normally appears black since little of the light entering the eye is reflected back. The diameter of the pupil is small under high illumination, but expands when illumination is low to allow more light to enter.

The retina owes its photosensitivity to a mosaic of light sensitive cells known as *rods* and *cones*, which derive their names from their physical shape. There are about 6 million cone cells, 120 million rod cells and 1 million nerve fibres distributed across the retina. It is the rods and cones that translate the optical image into a pattern of nerve activity that is transmitted to the brain by the fibres in the optic nerve. At low levels of illumination only the rod cells are active and a type of vision known as *scotopic* vision operates, while at medium and high illumination levels only the cone cells are active, and this is gives rise to *photopic* vision. Only one type of rod-shaped cell is present in the eye. The rods provide, essentially, a monochromatic view of the world, allowing perception only of lightness and darkness. The sensitivity of rods to light depends on the presence of a photosensitive pigment known as *rhodopsin*, which consists chemically of the carotenoid retinal bonded to the protein opsin. Rhodopsin is continuously generated in the eye and is also destroyed by bleaching on exposure to light. At low levels of illumination (night or dark-adapted vision), this rate of bleaching is low and thus there is sufficient rhodopsin present for the rods to be sensitive to the small amounts of light. At high levels of illumination, however, the rate of bleaching is high so that only a small amount of rhodopsin is present and the rods consequently have low sensitivity to light. At these higher levels of illumination, it is only the cone cells that are sensitive. The cones provide us with full colour vision as well as the ability to perceive lightness and darkness. The sensitivity of cones to light depends on the presence of the photosensitive pigment *iodopsin*, which is retained up to high levels of illumination. Thus, in normal daylight when the rods are inactive, vision is provided virtually entirely by the response of the cone cells. Under ideal conditions, a normal observer can distinguish about 10 million separate colours. Three separate types of cone cells have been identified in the eye and our ability to distinguish colours is associated with the fact that each of the three types is sensitive to light of a particular range of wavelengths. The three types of cone cell have been classified as long, medium and short, corresponding to the wavelength of maximum response of each type. Short cones are most sensitive to blue light, the maximum response being at a wavelength of about 440 nm. Medium cones are most sensitive to green light, the maximum response being at about wavelength of about 545 nm. Long cones are most sensitive to red light, the maximum response being at about 585 nm. The specific colour sensation perceived by the eye is governed by the response of these three types of cone cells to the particular wavelength profile with which they are interacting.

THE CAUSES OF COLOUR

It is commonly stated that there are fifteen specific causes of colour, arising from a variety of physical and chemical mechanisms. These mechanisms may be collected into five groups.

(a) Colour from simple excitations: colour from gas excitation (*e.g.* vapour lamps, neon signs), and colour from vibrations and rotations (*e.g.* ice, halogens);

(b) colour from ligand field effects: colour from transition metal compounds and from transition metal impurities;

(c) colour from molecular orbitals: colour from organic compounds and from charge transfer;

(d) colour from band theory: colour in metals, in semiconductors, in doped semiconductors and from colour centres;

(e) colour from geometrical and physical optics: colour from dispersion, scattering, interference and diffraction.

This book is focused on the industrially important organic dyes and pigments and, to a certain extent, inorganic pigments and thus deals almost exclusively with colour generated by the mechanisms described by group (c).

THE INTERACTION OF LIGHT WITH OBJECTS

The most obvious requirement of a dye or pigment to be useful in its applications is that it must have an appropriate colour. Of the many ways in which light can interact with objects, the two most important from the point of view of their influence on colour are *absorption* and *scattering*. Absorption is the process by which radiant energy is utilised to raise molecules in the object to higher energy states. Scattering is the interaction by which light is re-directed as a result of multiple refractions and reflections. In general, if only absorption is involved when light interacts with an object, then the object will appear transparent as the light that is not absorbed is transmitted through the object. If there are scattering centres present, the object will appear either translucent or opaque, depending on the degree of scattering, as light is reflected back to the observer.

Electronic spectroscopy, often referred to as UV/visible spectroscopy, is a useful instrumental technique for characterising the colours of dyes and pigments. These spectra may be obtained from appropriate samples either in transmission (absorption) or reflection mode. UV/visible absorption spectra of dyes in solution, such as that illustrated in Figure 2.3, provide important information to enable relationships between the colour and the molecular structure of the dyes to be developed.

A dye in solution owes its colour to the selective absorption by dye molecules of certain wavelengths of visible light. The remaining wavelengths of light are transmitted, thus giving rise to the observed colour. The absorption of visible light energy by the molecule promotes electrons in the molecule from a low energy state, or *ground state*, to a higher energy state, or *excited state*. The dye molecule is therefore said to undergo an electronic transition during this excitation process. The

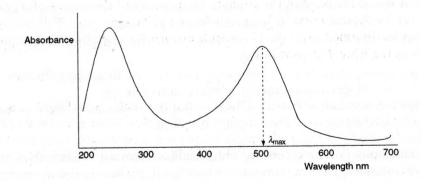

Figure 2.3   *UV/visible absorption spectrum of a typical red dye in solution*

energy difference, ΔE, between the electronic ground state and the electronic excited state is given by Planck's relationship

$$\Delta E = h\nu$$

where h is a constant (Planck's constant) and ν is the frequency of light absorbed. Alternatively, the relationship may be expressed as

$$\Delta E = hc/\lambda$$

where c is the velocity of light (also a constant) and λ is the wavelength of light absorbed. Thus there is an inverse relationship between the energy difference between the ground and excited states of the dye and the wavelength of light that it absorbs. As a consequence, for example, a yellow dye, which absorbs short wavelength (blue) light, requires a higher excitation energy than, say, a red dye which absorbs longer wavelength (bluish-green) light (Table 2.1).

There are a number of ways of describing in scientific terms the characteristics of a particular colour. One method which is especially useful for the purposes of relating the colour of a dye to its UV/visible spectrum in solution is to define the colour in terms of the three attributes: hue (or shade), strength (or intensity) and brightness. The hue of a dye is determined essentially by the absorbed wavelengths of light, and so it may be characterised to a reasonable approximation, at least in those cases where there is a single visible absorption band, by the λ_{max} value obtained from the UV/visible spectrum. A shift of the absorption band towards longer wavelengths (*i.e.* a change of hue in the direction yellow→orange→red→violet→blue→green), for example as a result of a structural change in a dye molecule, is referred to as a *bathochromic* shift. The reverse effect, a shift towards shorter absorbed wavelengths, is described as a *hypsochromic* shift.

A useful measure of the strength or intensity of the colour of a dye is given by the *molar extinction coefficient* (ε) at its λ_{max} value. This quantity may be obtained from the UV/visible absorption spectrum of the dye by using the Beer–Lambert law, *i.e.*

$$A = \varepsilon cl$$

where A is the absorbance of the dye at a particular wavelength, ε is the molar extinction coefficient at that wavelength, c is the concentration of the dye and l is the path length of the cell (commonly 1 cm) used for measurement of the spectrum. The Beer–Lambert law is obeyed by most dyes in solution at low concentrations, although when dyes show molecular aggregation effects in solution, deviations from the law may be encountered. However, since the colour strength of a dye is more correctly related to the area under the absorption band, it is important to treat its relationship with the molar extinction coefficient as qualitative and dependent to a certain extent on the shape of the absorption curve.

The third attribute, brightness, may be described in various other ways, for example as brilliance or vividness. This characteristic of the colour depends on the absence of wavelengths of transmitted light other than those of the hue concerned. Electronic absorption bands of molecular compounds are not infinitely narrow because they are broadened by the superimposition of numerous vibrational energy levels on both the ground and excited electronic states. Brightness of colour is characterised, in terms of the UV/visible spectrum, by the shape of the absorption band. Dyes which exhibit bright colours show narrow absorption bands, whereas broad absorption bands are characteristic of dull colours, such as browns, navy blues and olive greens.

Visible reflectance spectroscopy is used routinely to measure the colour of opaque objects such as textile fabrics, paint films and plastics for purposes such as colour matching and dye and pigment recipe prediction. In many ways, this technique may be considered as complementary to the use of visible absorption spectroscopy for the measurement of transparent dye solutions. Reflectance spectra of typical red, green and blue surfaces are shown in Figure 2.4. The spectrum of the red surface, for example, shows low reflectance (high absorption) in the 400–500 nm (blue) and 500–600 nm (green) ranges, and high reflectance of the red wavelengths (600–700 nm).

When colour is assessed on the basis of reflectance measurements, it is common to consider the three relevant attributes of perception of colour as hue, chroma (or saturation) which is the 'colourfulness' or richness of the colour, and lightness, which refers to the amount of reflected light. These three attributes may be described using the concept of colour space, which shows the relationships of colours to one another and which illustrates the three-dimensional nature of colour, as portrayed in Figure

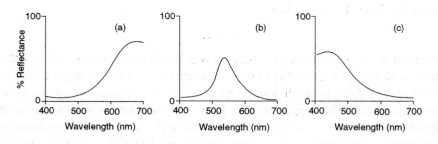

Figure 2.4   *Visible reflectance spectra of (a) red, (b) green and (c) blue surfaces*

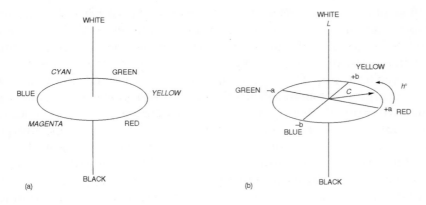

Figure 2.5   *(a)The concept of colour space; (b) LAB colour space*

2.5(a). The hue of a particular colour is represented in a colour circle. The three additive primaries, red, green and blue are equally spaced around the colour circle. The three subtractive primaries, yellow, magenta and cyan are located between the pairs of additive primaries from which they are obtained by mixing. The second attribute, chroma, increases with distance from the centre of the circle. The third attribute, lightness, requires a third dimension that is at right angles to the plane of the colour circle. The achromatic colours, white and black, are located at either extreme of the lightness scale. Mathematical approaches which make use of the concept of colour space for colour measurement and specification are now well known. In one of the most important of these approaches, the CIELAB equation for the measurement of colour differences makes use of the visually uniform LAB space, which is illustrated in Figure 2.5(b). Lightness, L and chroma, C, are quantified as illustrated on the diagram. Hue is described by the hue angle, $h°$. Colour measurement, and its mathematical basis, generally referred to as colour physics, is a well-developed and well-documented science and is not considered in further detail here.

Fluorescence and Phosphorescence

Most dyes and pigments owe their colour to the selective absorption of incident light. In some compounds, colour can also be observed as a result of the emission of visible light of specific wavelengths. These compounds are referred to as *luminescent*. The most commonly encountered luminescent effects are fluorescence and phosphorescence. The transitions which can occur in a molecule exhibiting either fluorescence or phosphorescence are illustrated in Figure 2.6. When the molecule absorbs light it is excited from the lowest vibrational level in its ground state (S_0) to a range of vibrational levels in the singlet first excited state (S_1^*). In the case of luminescent organic molecules, this is generally a $\pi-\pi^*$ electronic transition. During the time the molecule spends in the excited state, energy is dissipated from the higher vibrational levels, and the lowest vibrational level is attained. Fluorescence occurs if the molecule then emits light as it reverts from this level to various vibrational levels in the ground state. Non-radiative processes, the most important of which is generally collisional deactivation, also gives rise to dissipation of energy from the excited state. As a result, there will be a reduction in the intensity of fluorescence and in many cases it will be absent altogether. Another process which may occur is intersystem crossing to a triplet state. Emission of light from the triplet state is termed phosphorescence, a phenomenon which is longer-lived than fluorescence. As a consequence of the loss of vibrational energy in the excited state, fluorescent emission occurs at longer wavelengths than absorption, the difference between the wavelengths of maximum emission and maximum absorption for a fluorescent compound being referred to as the Stokes' shift. In the case of a fluorescent dye, the overall visual effect from the dye, whether in solution or when incorporated into a textile fabric, plastic material or surface coating, originates from the colour due to the selective absorption of light supplemented by the colour due to the light emitted. This gives the dye its particularly visual brilliance. Another type of compound which makes use of light-emitting properties is fluorescent brightening agents (FBAs). These are compounds, structurally closely related to fluorescent dyes,

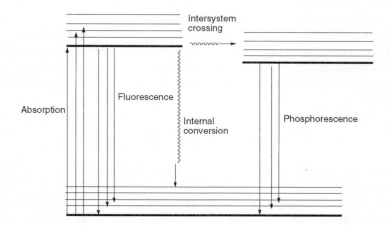

Figure 2.6   *Energy transitions in fluorescent and phosphorescent molecules*

that absorb light in the UV region of the spectrum and re-emit the energy at the lower (blue) end of the visible spectrum. When incorporated into a white substrate, such as a textile fabric or a plastic article, FBAs provide a particularly appealing bluish cast. One of the most important uses of FBAs is in washing powders to impart a bluish whiteness to washed fabrics.

DYES AND PIGMENTS

Colour may be introduced into manufactured articles, for example textiles and plastics, or into a range of other application media, for example paints and printing inks, for a variety of reasons, but most commonly the purpose is to enhance the appearance and attractiveness of a product and improve its market appeal. Indeed it is often the colour which first attracts us to a particular article. The desired colour is generally achieved by the incorporation into the product of coloured compounds referred to as dyes and pigments. The term *colorant* is frequently used to encompass both types of colouring materials. For an appreciation of the chemistry of colour application it is of fundamental importance that the distinction between dyes and pigments as quite different types of colouring materials is made. Dyes and pigments are both commonly supplied by the manufacturers as coloured powders. Indeed, as the discussion of their molecular structures contained in subsequent chapters of this book will illustrate, the two groups of colouring materials may often be quite similar chemically. However, they are distinctly different in their properties and especially in the way they are used. Dyes and pigments are distinguished on the basis of their solubility characteristics: essentially, dyes are soluble, pigments are insoluble.

The traditional use of dyes is in the coloration of textiles, a topic covered in considerable depth in Chapters 7 and 8. Dyes are almost invariably applied to the textile materials from an aqueous medium, so that they are generally required to dissolve in water. Frequently, as is the case for example with acid dyes, direct dyes, cationic dyes and reactive dyes, they dissolve completely and very readily in water. This is not true, however, of every application class of textile dye. Disperse dyes for polyester fibres, for example, are only sparingly soluble in water and are applied as a fine aqueous dispersion. Vat dyes, an important application class of dyes for cellulosic fibres, are completely insoluble materials but they are converted by a chemical reduction process into a water-soluble form that may then be applied to the fibre. There is also a wide range of non-textile applications of dyes, many of which have emerged in recent years as a result of developments in the electronic and reprographic industries (see Chapter 10). For many of these applications, solubility in specific organic solvents rather than in water is of importance.

In contrast, pigments are colouring materials that are required to be completely insoluble in the medium into which they are incorporated. The principal traditional applications of pigments are in paints, printing inks and plastics, although they are also used more widely, for example, in the coloration of building materials, such as concrete and cement, and in ceramics and glass. The chemistry of pigments and their application is discussed in more detail in Chapter 9. Pigments are applied into a medium by a dispersion process, which reduces the clusters of solid particles into a more finely-divided form, but they do not dissolve in the medium. They remain as solid particles held in place mechanically, usually in a matrix of a solid polymer. A further distinction between dyes and pigments is that while dye molecules are designed to be attracted strongly to the polymer molecules which make up the textile fibre, pigment molecules are not required to exhibit such affinity for their medium. Pigment molecules are, however, attracted strongly to one another in their solid crystal lattice structure in order to resist dissolving in solvents.

CLASSIFICATION OF COLORANTS

Colorants may be classified usefully in two separate ways, either according to their chemical structure or according to the method of application. The most important reference work dealing with the classification of dyes and pigments is the *Colour Index*, a publication produced by the Society of Dyers and Colourists, Bradford, England. This series of volumes provides a comprehensive listing of the known commercial dyes and pigments and is updated on a regular basis. Each colorant is given a C. I. Generic Name, which incorporates its application class, the hue and a number, which simply reflects the chronological order in which the colorants were introduced commercially. It is extremely useful that this system of nomenclature for dyes and pigments is more or less universally accepted by all those involved in their manufacture and application, and so it is frequently used throughout this book. The *Colour Index* provides useful information for each dye and pigment on the methods of application and on the range of fastness properties. The volumes also list the companies that manufacture each of the products, together with trade names, and give the appropriate chemical constitutions where these have been disclosed by the manufacturer.

In the chemical classification method, colorants are grouped according to certain common chemical structural features. The most important organic dyes and pigments, in roughly decreasing order of importance, belong to the azo (–N=N–), carbonyl (C=O) (including anthraquinones), phthalocyanine, arylcarbonium ion (including triphenylmethines), sulfur, polymethine and nitro chemical classes. A discussion of the principal structural characteristics of each of these chemical classes, together with a discussion of the major synthetic strategies used in the manufacture of these chemical groups of organic colorants, may be found in Chapters 3–6. Added to this, Chapter 9 contains a discussion of the most important inorganic pigments, a group of colorants that has no counterpart in dye chemistry.

To the textile dyer whose role it is to apply colour to a particular textile fibre, the classification of dyes according to the method of application is arguably of greater interest than the chemical classification. Dye and pigment molecules are carefully designed to ensure that they have a set of properties that are appropriate to their particular application. Obvious requirements for both types of colorant are that they must possess the desired colours, in terms of hue, strength and brightness, and an appropriate range of fastness properties. Fastness properties refer to the ability of a dye or pigment to resist colour change when exposed to certain conditions, such as to light, weathering, heat, washing, solvents or to chemical agencies such as acids and alkalis. For textile applications, dye molecules are designed so that they are attracted strongly to the molecules of the fibre to which they are applied. The chemical and physical nature of the different types of textile fibres, both natural and synthetic, require that the dyes used, in each case, have an appropriate set of chemical features to promote affinity for the particular fibre concerned. The chemical principles of the most important application classes of textile dyes are described in Chapters 7 and 8. The application classes discussed include acid dyes, mordant dyes and premetallised dyes for protein fibres, direct dyes, reactive dyes and vat dyes for cellulosic fibres, disperse dyes for polyester and basic (cationic) dyes for acrylic fibres. In contrast to textile dyes, pigments tend to be versatile colouring materials requiring less tailoring to individual applications. With pigments therefore, classification according to application is relatively unimportant. Pigments are simply designed to resist dissolving in solvents with which they may come into contact in paint, printing ink or plastics applications.

There is a third method proposed for classifying colorants which is in terms of the mechanism of the electronic excitation process. According to this method, organic colorants may be classified as donor–acceptor, polyene, cyanine or n–π^* chromogens. While this method of classification is undoubtedly of importance theoretically, it is arguably of lesser practical importance, since the vast majority of commercial organic dyes and pigments for traditional applications belong to the donor–acceptor category. The mechanism of the electronic excitation process, an understanding of which is fundamental in establishing relationships between the colour and the molecular constitution of dyes, is discussed in the following sections of this chapter.

COLOUR AND CONSTITUTION

Since the discovery of the first synthetic dyes in the mid-19th century, chemists have been intrigued by the relationship between the colour of a dye and its molecular structure. Since these early days, the subject has been of special academic interest to those fascinated by the origin of colour in organic molecules. In addition, an understanding of colour and constitution relationships has always been of critical importance in the design of new dyes. In the very early days of synthetic colour chemistry little was known about the structures of organic molecules. However, following Kekulé's proposal concerning the structure of benzene in 1865, organic chemistry made significant and rapid progress as a science and, almost immediately, theories concerning the influence of organic structures on the colour of molecules began to appear in the literature. One of the earliest observations of relevance was due to Graebe and Liebermann who, in 1867, noted that treatment of the dyes known at the time with reducing agents caused a rapid destruction of their colour. They concluded, with some justification, that the dyes were unsaturated compounds and that this unsaturation was destroyed by reduction.

Perhaps the most notable early contribution to the science of colour and constitution was due to Witt who, in 1876, proposed that dyes contain two types of group which are responsible for their colour. The first of these is referred to as the *chromophore*, which is defined as a group of atoms principally responsible for the colour of the dye. Secondly, there are the *auxochromes*, which he suggested were 'salt-forming' groups of atoms whose role, rather more loosely defined, was to provide an essential 'enhancement' of the colour. This terminology is still used to a certain extent today to provide a simple explanation of colour, although Witt's original suggestion that auxochromes were also essential for dyeing properties was quickly recognised as having less validity. A further notable contribution was made by Hewitt and Mitchell who first proposed in 1907 that conjugation is essential for the colour of a dye molecule. In 1928, this concept was incorporated by Dilthey and Witzinger in their refinement of Witt's theory of chromophores and auxochromes. They recognised that the chromophore is commonly an electron-withdrawing group, that auxochromes are usually electron-releasing groups and that they are linked to one another through a conjugated system. In essence, the concept of the donor–acceptor chromogen was born. Furthermore, it was observed that a bathochromic shift of the colour, *i.e.* a shift of the absorption band to longer wavelength, might be obtained by increasing the electron-withdrawing power of the chromophore, by increasing the electron-releasing power of the auxochromes and by extending the length of the conjugation.

The chromophore and auxochrome theory, which was first proposed more than 100 years ago, still retains some merit today as a simple method for explaining the origin of colour in dye molecules although it lacks rigorous theoretical justification. The most important chromophores, as defined in this way, are the azo (–N=N–), carbonyl (C=O), methine (–CH=) and nitro (NO_2) groups. Commonly-encountered auxochromes, groups that normally increase the intensity of the colour and shift the absorption to longer wavelengths of light, include hydroxyl (OH) and amino (NR_2) groups. The numerous examples of chemical structures which follow in later sections of this book will illustrate the many ways in which chromophores, auxochromes and conjugated aromatic systems, together with other structural features designed to confer particular application properties, are incorporated into dye and pigment molecules. The concept may be applied to most chemical classes of dye, including azo, carbonyl, methine and nitro dyes, but for some classes which are not of the donor–acceptor type, for example the phthalocyanines, it is less appropriate. Nowadays, modern theories of chemical bonding, based on either the valence-bond or the molecular orbital approaches, are capable of providing a much more sophisticated account of colour and constitution relationships.

The Valence-bond Approach to Colour and Constitution

The valence-bond (or resonance) approach to bonding in organic molecules is a particularly useful approach to explaining the properties of aromatic compounds. The approach involves postulating a series of organic structures that represent a particular compound in each of which the electrons are localised in bonds between atoms. These structures are referred to as canonical, or resonance, forms. The individual structures do not have a separate existence, but rather each makes a contribution to the overall structure of the molecule, which is considered to be a resonance hybrid of the contributing forms. Arguably the simplest example of resonance is benzene which, to a first approximation, may be considered as a resonance hybrid of the two Kekulé structures as shown in Figure 2.7. It was Bury who, in 1935, first highlighted the relationship between

Figure 2.7   *Valence-bond (resonance) approach to the structure of benzene*

resonance and the colour of a dye, noting that the more resonance structures, of comparable energy, which could be drawn for a particular dye, the more bathochromic were the dyes.

The valence-bond approach may be used to provide a qualitative account of the λ_{max} values, and hence the hues, of many dyes, particularly those of the donor–acceptor chromogen type. The use of this approach to rationalise differences in colour is illustrated in this section with reference to a series of dyes which may be envisaged as being derived from azobenzene, although in principle the method may be used to account for the colours of a much wider range of chemical classes of dye, including anthraquinones (see Chapter 4), polymethines and nitro dyes.

The valence-bond approach to colour and constitution requires that certain assumptions be made concerning the structures of the electronic ground state and of the electronic first excited state of the dye molecules. Invariably, a dye molecule may be represented as a resonance hybrid of a large number of resonance forms, some of which are 'neutral' or normal Kekulé-type structures, and some of which involve charge-separation, particularly involving electron release from the donor through to the acceptor groups. For the purpose of explaining the colour of dyes, a first assumption is made that the ground electronic state of the dye most closely resembles the most stable resonance forms, the normal Kekulé-type structures. A second assumption is that the first excited state of the dye more closely resembles the less stable, charge-separated forms. The nature of these assumptions will be clarified by a consideration of the examples that follow. As a consequence of Planck's relationship ($\Delta E = hc/\lambda$), the wavelength at which the dye absorbs increases (a bathochromic shift) as the difference in energy between the ground state and the first excited state decreases. Essentially the valence-bond approach is used to account for these energy differences. Structural factors, both electronic and steric, which either stabilise or destabilise the first excited state relative to the ground state, are analysed to provide a qualitative explanation of colour. The assumptions which the approach makes concerning the structures of the ground and the first excited states are clearly approximations and cannot be rigorously justified. Nevertheless, evidence that the approximations are reasonable is provided by the fact that the method works well, at least in qualitative terms, in such a large number of cases. There are, however, numerous examples where the approach fails, no doubt due to the inadequacy of the assumptions for those cases.

Table 2.2 shows the λ_{max} values obtained from the UV/visible spectra, recorded in the same solvent, of a series of substituted azobenzenes, which may be considered as the basis of the simplest azo dyes. A number of observations may be made from the data given in Table 2.2. Azobenzene **15a**, the simplest aromatic azo compound, gives a λ_{max} value of 320 nm. This compound is only weakly coloured because its principal absorption is in the UV region. An electron-withdrawing group, such as the nitro group, on its own produces only a weakly bathochromic shift, so that compound **15b** (λ_{max} 332 nm) is still only weakly coloured. In contrast, an electron-releasing group in the *para*-position leads to a pronounced bathochromic shift, the magnitude of which increases with increasing electron-releasing power of the group in question. The absorption bands of compounds **15c–e** are thus shifted into the visible region (λ_{max} 385, 407 and 415 nm respectively) and the compounds are reasonably intense yellows. Very large bathochromic shifts are provided when there is an electron-releasing group in one aromatic ring and an electron-withdrawing group in the other, *i.e.* a typical donor–acceptor chromogen, as in the case of compound **15f**, which is orange-red (λ_{max} 486 nm).

Valence-bond representations consisting of the most relevant canonical forms contributing towards the structures of compounds **15a–f** are illustrated in Figure 2.8. The arguments that follow illustrate how the valence-bond approach may be used to explain colour by rationalising the trends in λ_{max} values for the compounds given in Table 2.2. Using the approach, it is assumed that the ground state of azobenzene **15a**, the parent compound, most closely resembles the more stable traditional

Table 2.2 λ_{max} *Values for a series of substituted azobenzenes (* **15a–f** *)*

15

Compound	X	Y	λ_{max} (nm) in C_2H_5OH
15a	H	H	320
15b	NO$_2$	H	332
15c	H	NH$_2$	385
15d	H	NMe$_2$	407
15e	H	NEt$_2$	415
15f	NO$_2$	NEt$_2$	486

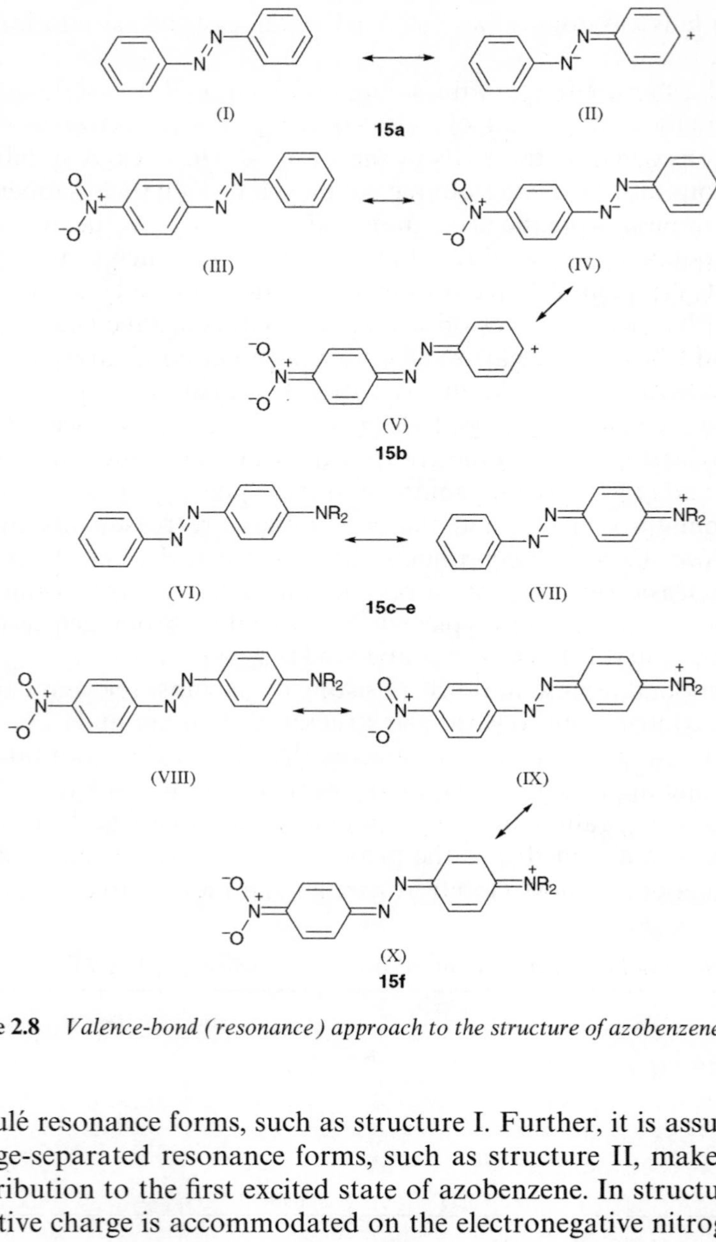

Figure 2.8 *Valence-bond (resonance) approach to the structure of azobenzenes* **15a–f**

Kekulé resonance forms, such as structure I. Further, it is assumed that charge-separated resonance forms, such as structure II, make a major contribution to the first excited state of azobenzene. In structure II, the negative charge is accommodated on the electronegative nitrogen atom of the azo group, a reasonably stable situation, but the structure also contains a carbocationic centre, a clear source of instability. The first excited state of azobenzene is therefore rather unstable, *i.e.* of high energy. In the case of 4-nitroazobenzene (**15b**), the carbocationic centre is still present as a destabilising feature in the first excited state (structures IV and V). There is, however, a marginal stabilisation of the first excited state in compound **15b** as a result of the additional contribution to the first excited state from canonical forms such as structure V, in which charge is delocalised onto the nitro group. Consequently, the energy difference between the ground and first excited states in compound **15b** becomes smaller than in azobenzene, **15a**, and a small bathochromic shift is observed as a result of the inverse relationship between energy difference and the absorbed wavelength.

The situation is significantly different with the 4-aminoazobenzenes **15c–e**. In the charge-separated structures which make the major contribution to the first excited states of these compounds (structure VII), donation of the lone pair from the amino nitrogen atom removes the carbocationic centre, which has a sextet of electrons in its valence shell, and the positive charge becomes accommodated on the nitrogen atom, deriving much greater stability from its full octet of electrons. There is thus a marked stabilisation of the first excited state, lowering its energy and leading to a pronounced bathochromic shift. The electron-releasing inductive effect of the *N*-alkyl groups in compounds **15d** and **15e** serves to increase the electron-donor power of the lone pair on the amino nitrogen atom compared with compound **15c** thus further stabilising the first excited state and causing a more pronounced bathochromic shift, the magnitude of which increases with increasing electron-releasing power of the alkyl groups. In the case of compound **15f**, typical of most aminoazobenzene dyes in which one aromatic ring contains electron-donor groups and the other electron-acceptor groups, the very strong bathochromicity is explained by a first excited state in which there is further stabilisation by charge delocalisation on to the nitro group, as a result of a contribution from structure X.

The spectral data for a further group of donor–acceptor aminoazobenzenes are given in Table 2.3. The valence-bond approach may be used to provide a good qualitative account of the data in the table. Some relevant resonance structures, which may be used to explain the λ_{max} values of cyano compounds **16a–d**, are shown in Figure 2.9.

Inspection of the spectral data for the isomeric cyano compounds **16a–c** demonstrates that the bathochromicity is most pronounced when the donor and acceptor groups are conjugated with one another, *i.e. ortho* or *para* to the azo group, allowing the full mesomeric interaction between the amino and cyano groups in the first excited states to operate, as illustrated for compounds **16a** and **16c** in structures XI and XII respectively. In contrast, the presence of a cyano group in the *meta* position gives rise to a less pronounced bathochromic effect (compound **16b**) as only the inductive effect of the cyano group is in operation. Increasing the number

Table 2.3 λ_{max} *values for a series of donor–acceptor substituted azobenzenes (*16a–f*)*

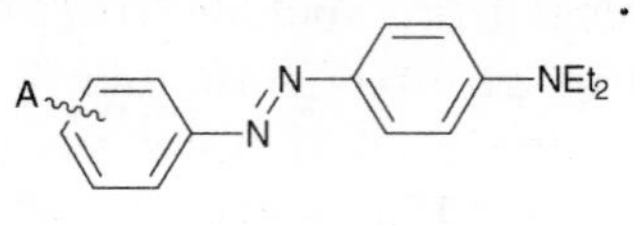

Compound	Acceptor substituents (A)	λ_{max} (nm) in C_2H_5OH
16a	*o*-CN	462
16b	*m*-CN	446
16c	*p*-CN	466
16d	*o*-, *o*′-, *p*-tricyano	562
16e	*o*-NO$_2$	462
16f	*p*-NO$_2$	486

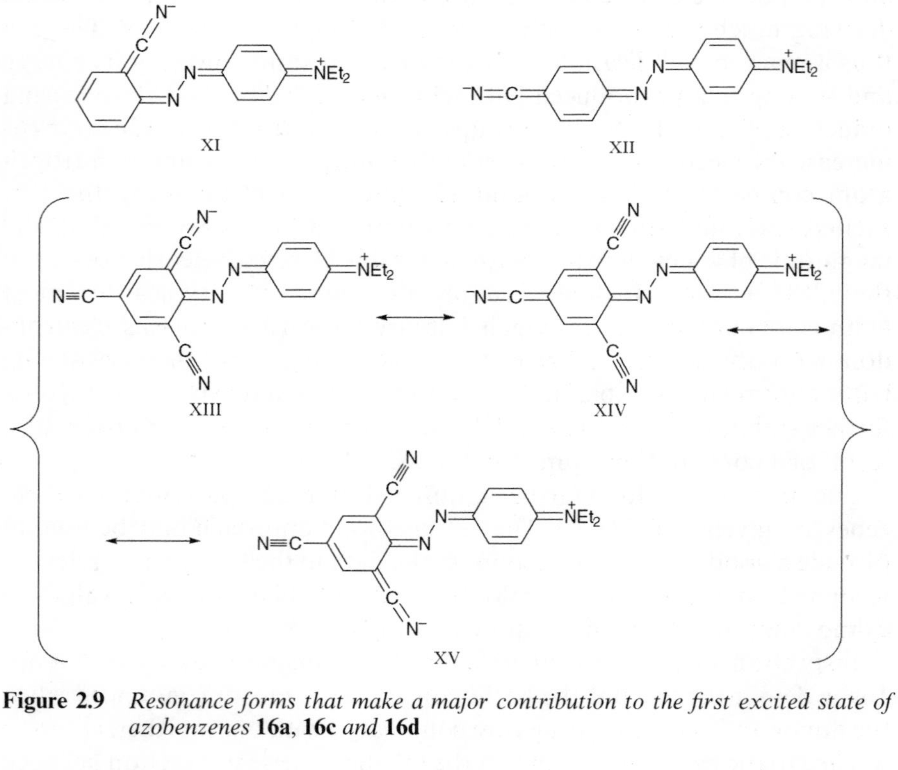

Figure 2.9 *Resonance forms that make a major contribution to the first excited state of azobenzenes* **16a**, **16c** *and* **16d**

of electron-withdrawing groups in the acceptor ring increases the bathochromicity, as illustrated by the tricyano compound **16d**. The bathochromicity of this dye (λ_{max} 562 nm) is explained by the extensive stabilisation of the first excited state as a result of resonance involving structures XIII–XV in which the negative charge is accommodated on the nitrogen atoms of each of the three cyano groups in turn.

The colour of dyes may be affected by steric as well as electronic effects. For example, a comparison of isomers **16e** and **16f** shows that an *o*-nitro group produces a significantly smaller bathochromic shift than a *p*-nitro group. On the basis of electronic effects alone, delocalisation of charge on to the oxygen atom of the *o*-nitro group, as illustrated by the valence-bond representation of the structure of compound **16e** given in Figure 2.10, might be expected to give rise to a bathochromic shift similar to that given by a *p*-nitro group. The reason for the differences arises from the fact that while compound **16f** is a planar molecule, steric congestion forces compound **16e** to adopt a non-planar conformation. This happens because the *o*-nitro group clashes sterically with the lone pair of electrons on one of the azo nitrogen atoms and is thus forced to rotate out of a planar conformation. It may be envisaged that this rotation takes place about the bond between the carbon atom of the aromatic ring and the nitrogen atom of the NO_2 group. As Figure 2.10 shows, the C–N bond in question is of higher bond order in the first excited state (represented approximately by structure XVII), but of low bond order in the ground state (similarly represented by structure XVI). Since rotation about a double bond requires more energy than rotation about a single bond, the first excited state of compound **16e** is destabilised relative to its ground state, with a consequent reduction in the bathochromic effect. When it is desirable to have electron-withdrawing substituents in the positions *ortho* to the azo group to maximise the bathochromicity, cyano groups are commonly preferred. Their linear, rod-like shape minimises the steric clash with the lone pair on the azo nitrogen atoms and allows a more planar conformation to be adopted.

The absorption band of aminoazobenzenes may be shifted to very long wavelengths to produce blue dyes by incorporating a number of electron-withdrawing groups into the acceptor ring and a number of electron-releasing groups into the donor ring. An example of such a dye is compound **17** which gives a λ_{max} of 608 nm in ethanol as a result of the three electron-withdrawing groups (two nitro, one bromo) in one ring and the three electron-releasing groups (dialkylamino, acylamino and

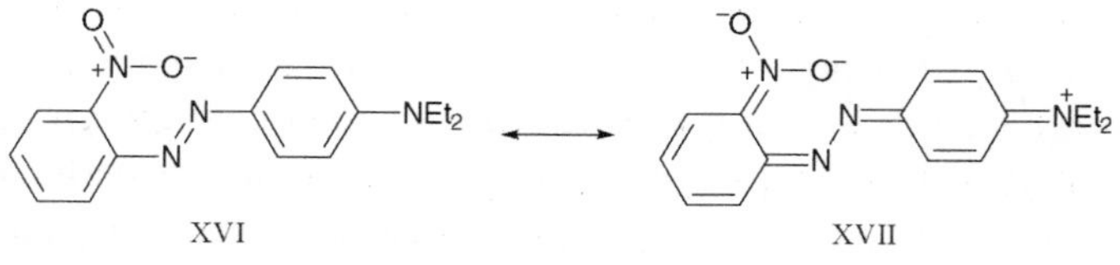

XVI   XVII

Figure 2.10 *Valence-bond (resonance) approach to the structure of compound* **16e**.

methoxy) in the other. Alternatively, particularly bathochromic azo dyes
that are structurally analogous to the aminoazobenzenes may be obtained by replacement of the carbocyclic acceptor ring with a five-membered aromatic heterocyclic ring. Examples of such dyes are provided by
C. I. Disperse Blue 339, **18**, a bright blue dye which contains a thiazole
ring and gives a λ_{max} of 590 nm and thiophene derivative **19** which gives a
λ_{max} of 614 nm.

A number of reasons for the particular bathochromicity of heterocyclic
azo dyes of this type may be postulated from the valence-bond approach.
Some important resonance forms which may contribute to the structure
of the first electronic excited state of dye **18** are illustrated in Figure 2.11
The first reason involves a consideration of aromatic character. It is
evident that, when the structures which are considered to make a major
contribution to the first excited states of azo dyes are compared with
those of the respective ground states, electronic excitation causes a loss of
aromatic character, *i.e.* a loss of resonance stabilisation energy. It is
generally accepted that five-membered ring heterocyclic systems of this
type are less aromatic than benzene derivatives. The loss of resonance
stabilisation energy on electronic excitation of dyes such as **18** is therefore
less than in the corresponding carbocyclic systems and, as a consequence,
the difference in energy between the ground and excited states is less.

Steric effects also play a part in the explanation. For example, the
contribution to the first excited state from structure XX (Figure 2.11)
illustrates that the heterocyclic nitrogen atom can effectively act as an
electron-withdrawing *ortho*-substituent with no associated steric effect
that might otherwise tend to reduce the bathochromicity. Also, in the case
of five-membered heterocyclic rings which contain *ortho*-substituents,
such as the *o*-nitro group in compound **19**, the steric interaction between
the substituent and the lone pair on the azo nitrogen atoms is less than in
the case of six-membered rings, allowing the group to adopt a more

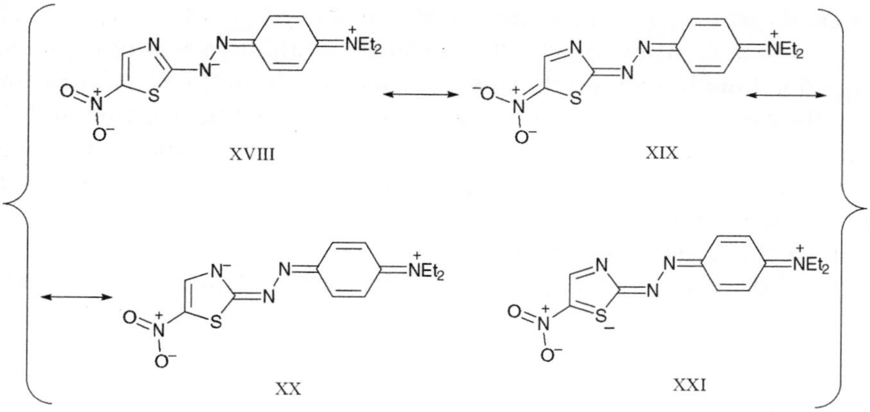

Figure 2.11 *Resonance forms that make a major contribution to the first excited state of heterocyclic azo dye* **18**

coplanar arrangement and hence maximise the bathochromic effect. This may be explained by a comparison of the geometrical arrangements. As illustrated in Figure 2.12, the azo group makes an angle of 126° with a five-membered ring, compared with a corresponding angle of 120° in the case of a six-membered ring, thus reducing the steric congestion between the oxygen atom of the nitro group and the lone pair on the azo nitrogen atom. A further explanation has been put forward to explain why sulfur heterocycles, in particular, appear in general to provide a pronounced bathochromic effect. It is argued, although it has to be said that there is no real consensus, that there is a contribution to the first excited state from structures such as **XXI** (Figure 2.11) in which there is valence shell expansion at sulfur. This is possible in principle as a consequence of the availability of vacant 3d-orbitals into which valence electrons may be donated, a situation that is not available in the case of nitrogen and oxygen heterocycles.

Compound **20** is noteworthy in that it was the first azo dye reported whose absorption band is shifted beyond the visible region into the

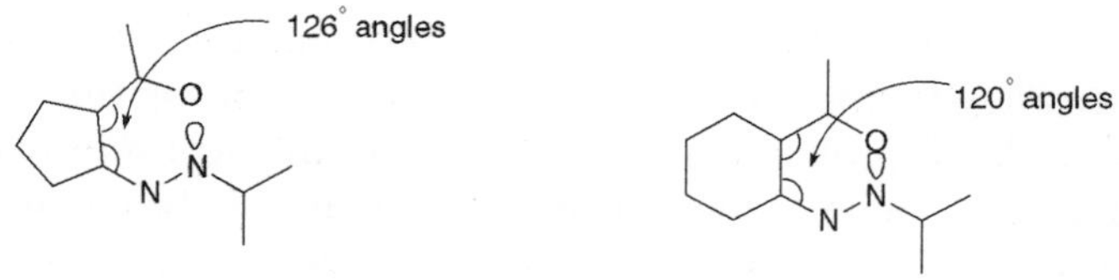

Figure 2.12 *Reduced steric congestion with substituents ortho to the azo group in the case of a five-membered ring compared with a six-membered ring*

near-infrared region of the spectrum, showing a λ_{max} of 778 nm in dichloromethane. The extreme bathochromicity of this dye may be explained by a combination of the effects discussed throughout this section, including the extended conjugation, the influence of the thiazole ring and the maximising of both the electron donor (dialkylamino, acylamino, methoxy) and electron accepting (cyano, sulfone, chloro) effects in appropriate parts of the molecule.

It can be seen that the valence-bond (resonance) approach may be used to provide a reasonable qualitative explanation of the λ_{max} values of a

20

wide variety of azo dyes. While the use of the approach has been exemplified in this chapter for a series of azo dyes, the argument may be extended successfully to a range of donor–acceptor type chromogens such as carbonyl (see Chapter 4 for an illustration of its application to some anthraquinone dyes), nitro and methine dyes. Nevertheless, the approach does have serious deficiencies. There are a number of examples where wrong predictions are made. For example, the order of bathochromicities of the *o-*, *m-* and *p-*aminoazobenzenes is wrongly predicted (see next section for a further discussion of this observation). Secondly, the approach cannot readily be used to account, even qualitatively, for the intensity of colour by addressing trends in molar extinction coefficients. Finally, and arguably most importantly, the method cannot at the present time be used quantitatively. Quantitative treatment of the light absorption properties of dyes has been made possible by developments in molecular orbital methods, which are discussed in the next section.

The Molecular Orbital Approach to Colour and Constitution

The potential value of the application of molecular orbital methods in colour chemistry is immense. In essence, the reason for this is that the methods enable, in principle, many of the light absorption properties of dyes, from a knowledge of their chemical structure, to be calculated with the aid of a computer. Thus, the colour properties of any dye whose structure may be drawn on paper may be predicted, with some expectation of accuracy, without the need to resort to devising a method for the synthesis of the dye in the laboratory. The value to the research chemist whose aim is the synthesis of new dyes with specific properties, perhaps for new applications, is obvious. The properties of a very large number of structures may be predicted in a short period of time using computational methods, and specific compounds for which interesting properties are predicted may be selected for synthesis and an evaluation of application performance. The advances in computer technology, both in terms of software and hardware, which have taken place in recent years and which appear set to continue into the future are certain to mean that the more sophisticated molecular orbital methods will become more and more accessible as a routine tool for the colour chemist. The concepts and mathematical basis of molecular theory are well documented, and this particular text makes little attempt to address the detail of these. The section which follows provides an outline of the basis of the molecular orbital approach to bonding, presenting an overview of the methods which are of particular value to the colour chemist at the present time, and the aspects of colour properties which the methods can address.

In molecular orbital theory, electrons are considered as a form of electromagnetic radiation, *i.e.* in terms of their wave nature rather than their particulate nature. A principle of fundamental importance to the theory is the quantum principle, which states that the electron can only exist in a fixed series of discrete energy states. An essential concept in quantum theory of relevance to colour chemistry is that electrons are contained in regions of high probability referred to as *orbitals*. The mathematics underlying molecular orbital theory was first formulated in 1925 by Schrödinger, the solution of whose equation gives a fixed number of values of E which are the energy states available to the electrons in a particular atom or molecule. Unfortunately because of the complexity of the mathematics, even with the computing power currently available, the equation may be solved exactly only for relatively simple atomic and molecular systems. However, a range of approximations may be made to obtain solutions to the equation. Coloured molecules have large molecular frameworks so that approximate methods for solution of the equation to give the required values of E (the energies of molecular orbitals) become even more essential. The simplest of these rely on a range of empirically or semi-empirically derived parameters

Molecular orbitals are considered to be generated by overlap of atomic orbitals. There are two types of overlap. Direct or 'end-on' overlap gives rise to σ-orbitals, either bonding types, the low energy orbitals which in the ground state of a molecule are occupied by two electrons, or the high energy antibonding (σ^*) orbitals which remain unoccupied, while π-molecular orbitals are obtained by indirect or 'sideways' overlap, for example from overlap of two singly occupied $2p_z$ atomic orbitals.

Dyes are usually organic molecules with extended conjugation, containing a framework of σ-bonds and an associated π-system. The lowest energy electronic transitions occur when an electron is promoted from an occupied π-orbital to an unoccupied π^* orbital. It is these $\pi-\pi^*$ transitions (rather than $\sigma-\sigma^*$ which are of much higher energy) that give rise to the absorption of organic dyes and pigments in the UV and visible regions of the spectrum.

The *Hückel molecular orbital* (HMO) method is one of the earliest and simplest molecular orbital methods and has proved to be a particularly appropriate method for calculations on conjugated molecules. In the HMO method, it is assumed that the molecular orbitals of a particular structure may be expressed as a linear combination (or sum) of the atomic orbitals present in the molecular system (the LCAO approximation). Essentially this represents a mathematical expression of the assumption that molecular orbitals are derived from overlap of atomic orbitals. The HMO method also assumes that in conjugated systems only the π-electrons are involved in the electronic transitions which take place when light is absorbed and that these transitions are unaffected by the framework of σ-bonds in the molecule (the $\sigma-\pi$ separation principle). This would at first sight appear to be a reasonable first approximation as in conjugated molecules it is generally assumed that the low energy $\pi-\pi^*$ transitions are principally responsible for UV and visible light absorption.

By calculating the energies of each molecular orbital, the **HMO** method may be used to provide electronic transition energies for the promotion of an electron from an occupied molecular orbital to a higher energy unoccupied orbital. These energies are obtained in terms of β, the bond resonance integral, which is treated as an empirical parameter and given a value by comparison of the calculated values with the experimental values obtained from UV/visible spectra. Two other sets of molecular parameters which may be calculated using the HMO method are the π-electron charge densities, Q, the measure of the π-electron charge localisation on each atom in the molecule, and the π-bond orders (P), the measure of the degree of π-overlap of atomic orbitals between each pair of atoms in the molecule. While the HMO method is simple and useful qualitatively, quantitative correlation with spectral data is only reasonable in a few special cases. The method has been shown to give a good correlation between experimental and calculated electronic transition energies with a series of aromatic hydrocarbons, and also with the set of linear polyenes, in the latter case provided that different bond β-values are used for the formal single and double bonds. In general, while the method appears to give reasonable predictions for hydrocarbons, it works much less well with molecules containing heteroatoms, such as O and N, and this severely restricts its usefulness as a tool for calculating the colour of dye molecules.

The molecular orbital method which has been most extensively and successfully applied to the calculation of the colour properties of dye molecules is the Pariser–Pople–Parr (PPP) approach. Like the HMO method, the PPP-MO method uses as its basic mathematical premise the LCAO approximation in combining atomic orbitals to form molecular orbitals. In addition, it retains the σ–π separation principle, essentially neglecting the influence of σ-electrons on the colour. The main way in which the PPP-MO method provides a refinement over the HMO method is by taking account of interelectronic interaction energies, and, in doing so, specifically includes molecular geometry, both of these features being ignored by the HMO method. The theoretical basis of the PPP-MO method is not discussed here, but an outline of the way in which a calculation of this type may be carried out follows.

The sequence of operations involved in a PPP-MO calculation is illustrated in the flow diagram shown in Figure 2.13. The method is illustrated for the case of 4-aminoazobenzene, **15c**. The first step involves devising a numbering system for all of the atoms that contribute to the π-system of the molecule. Secondly the total number of π-electrons in the

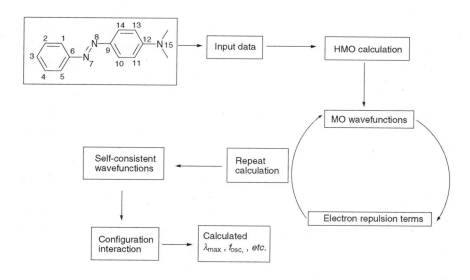

Figure 2.13 *The sequence of stages in a PPP-MO calculation for compound **15c**, for which an appropriate atomic numbering system is given*

molecule is indicated. In this case, there are 15 relevant atoms and 16 π-electrons, each atom contributing a single electron except for the amino nitrogen atom which donates its lone pair to the π-system. The molecular geometry is then specified in terms of all the interatomic distances and bond angles. Energy parameters, specifically valence state ionisation potentials (VSIP) and electron affinities (E_a), are assigned to all of the relevant atoms and bond resonance integrals values (β) are assigned to every pair of bonded atoms. The VSIP, E_a and β-values are treated as a set of semi-empirical parameters which are available from various literature sources and which are commonly modified to suit a particular type of molecule. Finally, to enable the calculations to be carried out the π-electron charge densities (Q) for all relevant atoms and the π-bond orders (P) for all bonded atoms are required. This feature of the PPP-MO method presents the problem that the values for Q and P for specific molecules are not known initially, but they may be obtained after the calculations have been carried out. The solution to this 'chicken and egg' problem is achieved by carrying out a preliminary HMO calculation, which gives an approximate set of Q and P values. These values may in turn be used to set up and carry out the PPP-MO calculation and as a result a new, improved set of values is obtained. The process is then repeated until two successive calculations give a consistent result. This iterative process is referred to as a self-consistent field (SCF) method. The procedure leads to a set of molecular orbital energies from which, in principle, electronic transition energies may be calculated. However, since the PPP-MO method uses a set of data based on the ground state structure of the molecule, it predicts ground state energies rather better than excited state energies. To correct for the fact that molecular orbital energies may change after excitation by promotion of an electron from a lower energy occupied molecular orbital to a higher energy unoccupied molecular orbital, and hence to give improved excited state data, a procedure known as configuration interaction (C.I.) is carried out as the last stage of the calculation.

The PPP-MO method is suitable for the treatment of large molecules, does not present major computing demands and programs are now routinely used as a tool to calculate the colour properties of dyes. Unlike the HMO method, it handles heteroatomic species well. The method has been remarkably successful in calculating λ_{max} values for a wide range of dyes from virtually all of the chemical classes. For example, the method provides a reasonably accurate account of substituent effects in the range of aminoazobenzene dyes, including compounds **15a–f** and **16a–f** which have been discussed in terms of the valence-bond approach in the previous section of this chapter.

Table 2.4 *Experimental and PPP-MO calculated electronic spectral data for azobenzene and the isomeric amino derivatives*

Compound	λ_{max} (nm) (expt.)	λ_{max} (nm) (PPP calc.)	$\varepsilon \times 10^{-3}$ $(l\ mol^{-1}\ cm^{-1})$	f_{osc} (PPP calc.)
Azobenzene, **15a**	320	344	21.0	1.21
2-Aminoazobenzene	414	424	6.5	0.67
3-Aminoazobenzene	412	415	1.3	0.17
4-Aminoazobenzene	387	392	24.5	1.31

Table 2.4 shows a comparison of the experimental and PPP-MO calculated electronic spectral data for azobenzene and the three isomeric monoamino derivatives. It is noteworthy that the *ortho* isomer is observed to be most bathochromic, while the *para* isomer is least bathochromic. From a consideration of the principles of the application of the valence-bond approach to colour described in the previous section, it might have been expected that the *ortho* and *para* isomers would be most bathochromic with the *meta* isomer least bathochromic. In contrast, the data contained in Table 2.4 demonstrate that the PPP-MO method is capable of correctly accounting for the relative bathochromicities of the amino isomers. It is clear, at least in this case, that the valence-bond method is inferior to the molecular orbital approach. An explanation for the failure of the valence-bond method to predict the order of bathochromicities of the o-, m- and p-aminoazobenzenes emerges from a consideration of the changes in π-electron charge densities on excitation calculated by the PPP-MO method, as illustrated in Figure 2.14.

The valence-bond representation of the ground and first excited states of dye **15c**, illustrated in Figure 2.8, would suggest that a decrease in charge density on the amino nitrogen atom and an increase in charge density on the β-nitrogen of the azo group would be observed on excitation. The changes in charge densities calculated by the PPP-MO (Figure 2.14) indeed predict these effects, but suggest that there is also a significant increase in the charge density of the α-nitrogen atom of the azo

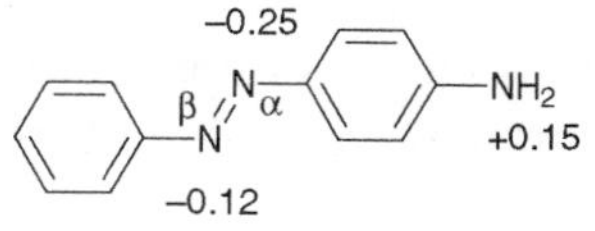

Figure 2.14 *Some π-electron charge density differences between the ground and first excited states calculated by the PPP-MO method for 4-aminoazobenzene,* **15c**.

group, an effect which may not be accounted for using the valence-bond approach. The results demonstrate that the valence-bond assumptions, particularly concerning the structure of the first excited state, are not wholly accurate and as a result it is perhaps not surprising that erroneous predictions are sometimes made.

Because molecular orbital methods such as the PPP approach are capable of calculating π-electron charge densities both in the electronic ground states and excited states of dye molecules, they are particularly helpful in providing information on the nature of the electronic excitation process, by identifying the donor and acceptor groups and quantifying the extent of the electron transfer. One use of this is that it allows the dipole moments of the ground and first excited state to be calculated, and this may be used to account for the influence of the nature of the solvent on the λ_{max} value of a dye, an effect referred to as *solvatochromism*. For example, if the dipole moment of a dye molecule is larger in the first excited state than in the ground state, then the effect of a more polar solvent will be to stabilise the first excited state more than the ground state. The consequence will be a bathochromic shift of the absorption maximum as the solvent polarity is increased, an effect known as positive solvatochromism. The reverse effect in which increasing solvent polarity causes a hypsochromic shift is perhaps not surprisingly referred to as negative solvatochromism.

The PPP-MO method is capable of calculating not only the magnitude of the dipole moment change on excitation, but it can also predict the direction of the electron transfer. The vector quantity that expresses the magnitude and direction of the electronic transition is referred to as the *transition dipole moment*. For example, the direction of the transition dipole moment of azo dye **15f** as calculated by the PPP-MO method is illustrated in Figure 2.15.

The direction of the transition moment is of practical consequence in dyes used in liquid crystal display systems. It is important for such applications that the direction of the transition moment is aligned with the molecular axis of the dye. Since this is the case with azo dye **15f**, the dye would appear to be a reasonable candidate as a liquid crystal display dye (see Chapter 10 for further discussion of this application of dyes).

The intensity of colour of a dye is dependent on the probability of the electronic transition. A familiar example of this principle is provided by

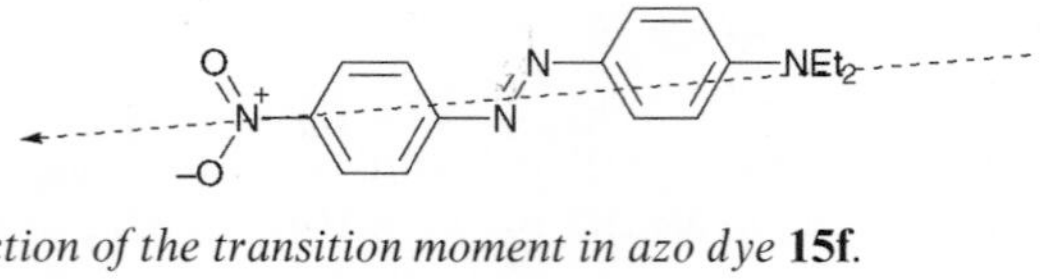

Figure 2.15   *Direction of the transition moment in azo dye* **15f**.

the colours due to transition metal ions in solution which are normally weak because the d–d transitions involved are 'forbidden', *i.e.* of low probability. In contrast the π–π^* transitions due to organic dyes, which involve considerable charge transfer in donor–acceptor chromogens, are highly probable and thus give rise to much more intense colours. While most of the research published on the application of molecular orbital methods, such as the PPP approach, centres on their application to the calculation of λ_{max} values, the ability to predict the tinctorial strength of a dye is arguably of greater practical value since it is directly related to the economic viability of the dye. If, for example a new dye has twice the colour strength of an existing dye, then the dyer need only use half the quantity of that dye to obtain a given colour. Provided that the new dye costs less than double that of the existing dye, it will therefore be more cost effective. The PPP-MO method is capable of providing a quantitative account of colour strength by calculating a quantity, f_{osc}, known as the oscillator strength. This parameter is given by the following equation:

$$f_{osc} = 4.703 \times 10^{29} \times M^2 / \lambda_M$$

where M is the transition dipole moment and λ_M is the mean wavelength of the absorption band. As the data given in Table 2.4 demonstrate, there is a reasonable correlation in qualitative terms between the calculated oscillator strengths and the experimental molar extinction coefficients, ε, for the series of azo dyes in question. In general, a reasonable correlation between f_{osc} and ε for broad classes of dyes may be achieved, although within specific classes the correlation is less good. This may well be due to the fact that PPP parameterisation has been optimised for correlation with λ_{max} values rather than molar extinction coefficients. It may be argued that the oscillator strength gives a better measure of colour intensity than the molar extinction coefficient (ε) as it expresses the area under the absorption curve, whereas ε is profoundly dependent on the shape of the curve. It is only valid to relate the ε values to the intensity of colour for a series of dyes if the curves are of similar shape.

Brightness of colour is expressed by the width of the visible absorption band. This bandwidth is determined by the distribution of vibrational energy levels superimposed on the electronic ground and excited state energy levels. Broadening of the absorption bands may be caused in a number of ways. For example, an increase in the number and spread of energies of bond vibrations will generally lead to broader absorption bands. This argument may be used to provide an explanation as to why the relatively simple structure of heterocyclic azo dyes such as compound **18** and **19** give brighter blue colours than the multi-substituted carbocyclic analogues such as compound **17** with its increased number of vibrational levels. Some of the brightest colours are provided by the phthalocyanine chemical class (see Chapter 5). The colour of these dyes and pigments owe their brightness in part to their rigid molecular structure both in the ground and excited states, both states showing similar geometry and little vibrational fine structure. Because it is related to the vibrational characteristics of the molecules rather than their electronic structure, it is at first sight difficult to envisage how the PPP-MO might be useful in calculating bandwidths. A solution to this problem has been provided by the application of an empirical extension of the PPP-MO method to the calculation of the Stokes' shifts of fluorescent molecules. Using the assumption that there is a simple relationship between absorption bandwidths and Stokes' shifts, both parameters being dependent on vibrational energy levels in the two electronic states, the method has been adapted, with some success, to the calculation of bandwidths.

The PPP-MO method has proved extremely successful for the prediction of a wide range of colour properties, and it is currently the most extensively used method for this purpose. It does have some deficiencies. For example, the method carries out its calculations based on π-electrons only and therefore cannot, except in a rather empirical way, account for some of the subtle effects of σ-electrons on colour. Among such effects commonly encountered are hydrogen bonding and steric hindrance. As more and more powerful computing facilities become accessible, there is clear evidence that colour chemists are turning their attention towards the use for colour prediction of more sophisticated molecular orbital techniques which take into account all valence electrons, such as the CNDO and ZINDO approaches, and in due course they may well prove to be the methods of choice. However, at the present time, it has not been established with absolute certainty that these methods will routinely provide superior colour prediction properties.

Chapter 3

Azo Dyes and Pigments

Azo dyes and pigments constitute by far the most important chemical class of commercial organic colorant. They account for around 60–70% of the dyes used in traditional textile applications (see Chapters 7 and 8) and they occupy a similarly prominent position in the range of classical organic pigments (see Chapter 9). Azo colorants, as the name implies, contain as their common structural feature the azo (–N=N–) linkage which is attached at either side to two sp^2 carbon atoms. Usually, although not exclusively, the azo group links two aromatic ring systems. The majority of the commercially important azo colorants contain a single azo group and are therefore referred to as monoazo dyes or pigments, but there are many which contain two (disazo), three (trisazo) or more such groups. In terms of their colour properties, azo colorants are capable of providing virtually a complete range of hues. There is no doubt though that they are significantly more important commercially in yellow, orange and red colours (*i.e.* absorbing at shorter wavelengths), than in blues and greens. However, as a result of relatively recent research, the range of longer wavelength absorbing azo dyes has been extended, leading to the emergence of significant numbers of commercially important blue azo dyes and there are even a few specifically-designed azo compounds which absorb in the near-infrared region of the spectrum (see Chapter 2 for a discussion of colour and constitution relationships in azo dyes). Azo colorants are capable of providing high intensity of colour, about twice that of the anthraquinones for example (see Chapter 4), and reasonably bright colours. They are capable of providing reasonable to very good technical properties, for example fastness to light, heat, water and other solvents, although in this respect they are often inferior to other chemical classes, for example carbonyl and phthalocyanine colorants, especially in terms of lightfastness.

Perhaps the prime reason for the commercial importance of azo colorants is that they are the most cost-effective of all the chemical classes of organic dyes and pigments. The reasons for this may be found in the nature of the processes used in their manufacture. The synthesis of azo colorants, which is discussed in some detail later in this chapter, brings together two organic components, a diazo component and a coupling component in a two-stage sequence of reactions known as diazotisation and azo coupling. The versatility of the chemistry involved in this synthetic sequence means that an immense number of azo colorants may be prepared and this accounts for the fact that they have been adapted structurally to meet the requirements of most colour applications. On an industrial scale, the processes are straightforward, making use of simple multi-purpose chemical plant. They are usually capable of production in high, often virtually quantitative, yields and the processes are carried out at or below ambient temperatures, thus presenting low energy requirements. The syntheses involve low cost, readily available commodity organic starting materials such as aromatic amines and phenols. The solvent in which the reactions are carried out is water, which offers obvious economic and environmental advantages over all other solvents. In fact, it is likely that in the future azo dyes are likely to assume even greater importance as some of the other chemical types, notably anthraquinones, become progressively less economic. This chapter contains a discussion of the fundamental structural chemistry of azo colorants, including a description of the types of isomerism that they can exhibit, and the principles of their synthesis. In the final section, the ability of azo yes to form metal complexes is discussed. Because of their prominence in most applications, numerous further examples of azo dyes and pigments will be encountered throughout this book.

ISOMERISM IN AZO DYES AND PIGMENTS

The structural chemistry of azo compounds is complicated by the possibilities of isomerism. There are two types of isomerism, which may commonly be encountered with certain azo compounds: geometrical isomerism and tautomerism.

Some simple azo compounds, because of restricted rotation about the (–N=N–) double bond, are capable of exhibiting geometrical isomerism. The geometrical isomerism of azobenzene, the simplest aromatic azo compound which may be considered as the parent system on which the structures of most azo colorants are based, is illustrated in Figure 3.1. The compound is only weakly coloured because it absorbs mainly in the UV region giving a $\lambda_{\max}$ value of 320 nm in solution in ethanol, a feature which may be attributed to the absence of auxochromes (see Chapter 2).

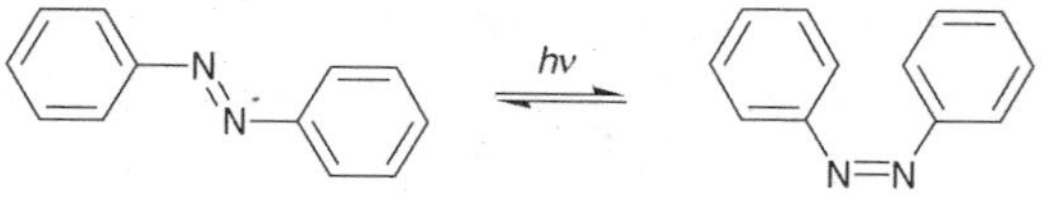

21a   **21b**

Figure 3.1 *The photo-induced geometrical isomerism of azobenzene*

The compound exists normally as the *trans* or (*E*)-isomer **21a**. This molecule is essentially planar both in the solid state and in solution, although in the gas phase there is evidence that it deviates from planarity. When irradiated with UV light, the (*E*)-isomer undergoes conversion substantially into the *cis* or (*Z*)-isomer **21b** which may be isolated as a pure compound. In darkness, the (*Z*)-isomer reverts thermally to the (*E*)-isomer which is thermodynamically more stable because of reduced steric congestion. Some early disperse dyes, which were relatively simple azobenzene derivatives introduced commercially initially for application to cellulose acetate fibres, were found to be prone to photochromism (formerly referred to as phototropy), a reversible light-induced colour change. C. I. Disperse Red 1 (**22**) is an example of a dye which has been observed, under certain circumstances, to give rise to this phenomenon.

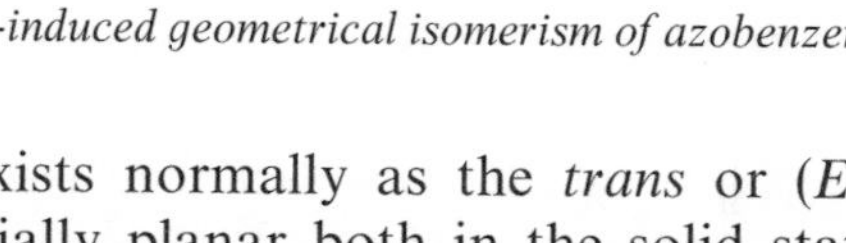

22

The phenomenon is explained by the colour changes associated with the *E/Z* isomerisation process, which may be initiated by exposure to light, the two geometrical isomers of the azo dyes having noticeably different colours. The undesirable effect is no longer encountered to any significant extent in the modern range of azo disperse dyes as the offending dyes were progressively replaced by more stable products. Curiously, there has been a relatively recent significant revival of interest in dyes which can exhibit controlled reversible photochromism for applications such as ophthalmic lenses, car sunroofs and optical data storage in which the colour change is utilised. Among the most promising photochromic dyes of this type are the spirooxazines (see Chapter 10 for further discussion).

Many commercial azo colorants contain a hydroxyl group *ortho* to the azo group. As illustrated in Figure 3.2, this gives rise to intramolecular hydrogen-bonding which further stabilises the (*E*)-isomer and effectively prevents its conversion into the (*Z*)-form.

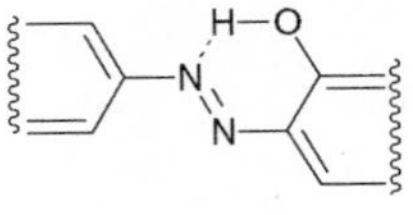

Figure 3.2 *Intramolecular hydrogen bonding in o-hydroxyazo compounds*

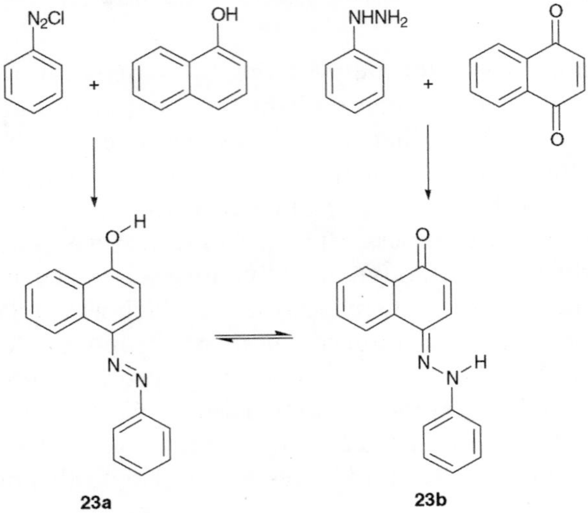

23a   **23b**

Scheme 3.1 *Reaction scheme which provides evidence for tautomerism in some hydroxyazo dyes*

Another important feature of azo compounds in which there is a hydroxy group conjugated with (*i.e. ortho* or *para* to) the azo group, and very many commercial azo dyes and pigments show this structural feature, is that they can exhibit tautomerism. The classical experimental evidence for hydroxyazo/ketohydrazone tautomerism in dyes of this type was provided by Zincke. In 1884, he carried out the reactions of benzenediazonium chloride with 1-naphthol and of phenylhydrazine with naphtho-1,4-quinone. It might have been expected that the former reaction would give hydroxyazo compound **23a**, and the latter would give the ketohydrazone tautomer **23b**. In the event, it was found that both reactions gave the same product, a tautomeric mixture of **23a** and **23b**, as illustrated in Scheme 3.1. These tautomeric forms may nowadays be readily identified by their distinctive UV/visible, IR and ^{1}H, ^{13}C and ^{15}N NMR spectral characteristics in solution, and by X-ray crystallography in the solid state. In terms of colour, the ketohydrazone isomers are usually bathochromic compared with the hydroxyazo forms, and give higher molar extinction coefficients.

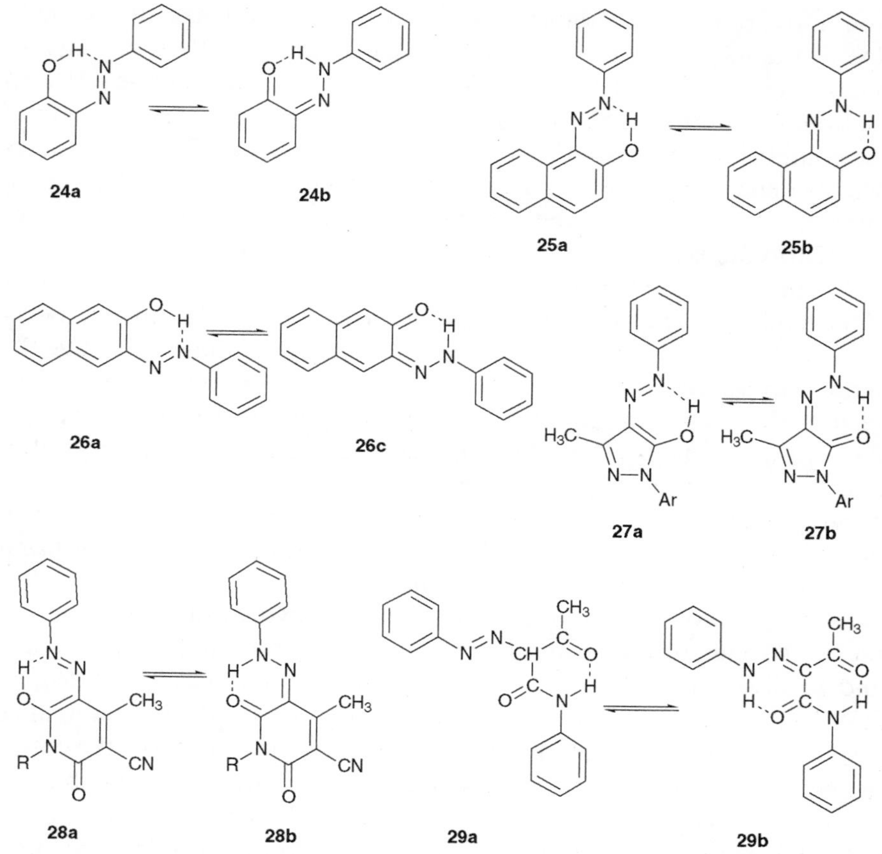

Figure 3.3 *Tautomeric forms of azo dyes* **24–29**

Figure 3.3 shows the tautomeric forms of a further range of hydroxyazo dyes. In solution, the tautomers exist in rapid equilibrium although commonly one or other of the tautomers is found to predominate to an extent that is dependent on their relative thermodynamic stability. In the case of 2-phenylazophenol, the azo isomer **24a** predominates over the hydrazone isomer **24b**. On the basis of a consideration solely of the summation of all of the theoretical bond energies in the molecule, it has been suggested that the hydrazone isomer would be expected to be more stable. However, this is outweighed by the considerably reduced reson-ance stabilisation energy in the hydrazone form, due to the loss of the aromatic character of one ring. In the case of hydroxyazonaphthalenes, 4-phenylazo-1-naphthol (**23**) and 1-phenylazo-2-naphthol (**25**), the reduction in the resonance stabilisation energy in the ketohydrazone forms is due to the loss of aromaticity of only one of the naphthalene rings. In relative terms, the reduction in energy is less than in the benzene series, so that the two tautomers become much closer in stability. The position of the equilibrium, particularly in the case of compound **23**, then becomes heavily dependent on the environment in which the dye finds itself, most importantly the nature of the solvent. Interestingly, 3-phenylazo-2-naphthol (**26**) exists exclusively in the azo form **26a**, because of the instability of hydrazone form **26b** in which there is complete loss of aromatic character of the naphthalene system. In the case of azo dyes derived from heterocyclic coupling components, such as the azopyrazolones **27** and the azopyridones **28**, and from β-ketoacid-derived coupling components, such as the azoacetoacetanilides **29**, the compounds exist exclusively in the hydrazone forms **27b**, **28b** and **29b** respectively. X-ray crystallographic structure determinations have been carried out on a wide range of azo pigments, and these compounds have been shown, without exception, to exist exclusively in the ketohydrazone form in the solid state (for a more detailed discussion of this see Chapter 9).

In *o*-hydroxyazo compounds, but not in *p*-hydroxyazo compounds, there is strong intramolecular hydrogen bonding both in the hydroxyazo and ketohydrazone forms, as illustrated for compounds **24–29** (Figure 3.3). A factor which contributes to the explanation for the predominance of the hydrazone isomer in many cases is that in this form the intramolecular hydrogen-bonding is significantly stronger than in the azo form. This may provide an explanation for the observations that in the case of 1-phenylazo-2-naphthol (**25**), the hydrazone form **25b** predominates while in the case of the isomeric 4-phenylazo-1-naphthol (**23**), in which there is no intramolecular hydrogen bonding, the two possible tautomers are of comparable stability. Intramolecular hydrogen bonding is a commonly encountered stabilising feature in a wide range of dyes and pigments of most chemical types, enhancing many useful technical properties. For example, lightfastness is usually improved. This is explained by a reduction in the electron density at the azo group due to hydrogen bonding, which decreases its sensitivity towards photochemical oxidation. Also, hydrogen bonding reduces the acidity of a hydroxyl group, giving the dye enhanced stability towards alkali treatments.

SYNTHESIS OF AZO DYES AND PIGMENTS

Textbooks in general organic chemistry will illustrate that there are many ways of synthesising azo compounds. However, almost without exception, azo dyes and pigments are made on an industrial scale by the same

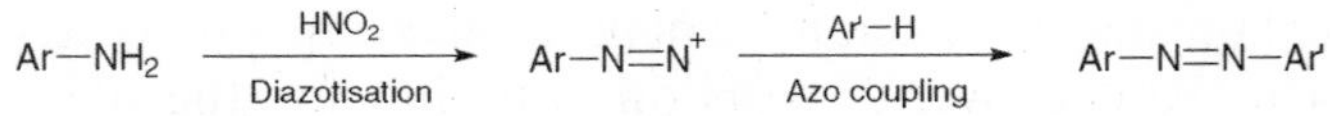

Scheme 3.2 *The synthesis of azo colorants*

two-stage reaction sequence: diazotisation and azo coupling, as illustrated in Scheme 3.2.

The first stage, diazotisation, involves the treatment of a primary aromatic amine, referred to as the diazo component, with sodium nitrite under conditions of controlled acidity and at relatively low temperatures to form a diazonium salt. In the second stage of the sequence, azo coupling, the relatively unstable diazonium salt thus formed is reacted with a coupling component, which may be a phenol, an aromatic amine or a β-ketoacid derivative, to form the azo dye or pigment. The next two sections of this chapter deal separately with these two reactions, with emphasis on the practical conditions used for the reactions and on the reaction mechanisms. This sequence of reactions provides an interesting illustration for students of organic chemistry of the ways in which the selection of the optimum practical conditions for the reactions is heavily influenced by consideration of the reaction mechanisms which are operating.

Diazotisation

Diazotisation, the first stage of azo dye and pigment synthesis, involves the treatment of a primary aromatic amine ($ArNH_2$), which may be carbocyclic or heterocyclic, with nitrous acid to form a diazonium salt ($ArN_2^+Cl^-$). Nitrous acid, HNO_2, is a rather unstable substance that decomposes relatively easily by dissociation into oxides of nitrogen. It is therefore usually generated in the reaction mixture as required by treating sodium nitrite, a stable species, with a strong acid. The mineral acid of choice for many diazotisations is hydrochloric acid. This is because the presence of the chloride ion can exert a catalytic effect on the reaction under appropriate conditions, thus enhancing the reaction rate. Most primary aromatic amines undergo diazotisation with little interference from the presence of other substituents, although they may influence the reaction conditions required. When the reaction conditions are carefully controlled, diazotisation usually proceeds smoothly and in virtually quantitative yield. It is of considerable industrial importance that the reaction may be carried out in water, the reaction solvent of choice for obvious economic and environmental reasons.

Diazotisation is always carried out under strongly acidic conditions, but control of the degree of acidity is of particular importance in ensuring smooth reaction. The overall reaction equation for the diazotisation reaction using sodium nitrite and hydrochloric acid may be given as:

$$ArNH_2 + NaNO_2 + 2HCl \rightarrow ArN_2^+Cl^- + H_2O$$

The reaction stoichiometry therefore requires the use of two moles of acid per mole of amine. However, for a number of reasons, a somewhat greater excess of acid is generally used. One reason is that highly acidic conditions favour the generation from nitrous acid of the reactive nitrosating species which are responsible for the reaction, a feature which will emerge from the discussion of the reaction mechanism later in this chapter. A second reason is that acidic conditions suppress the formation of triazines as side-products which may be formed as a result of N-coupling reactions between the diazonium salts and the aromatic amines from which they are formed. A practical reason for the use of acidic conditions is to convert the insoluble free amine ($ArNH_2$) into its water-soluble protonated form ($ArNH_3^+Cl^-$). However, too strongly acidic conditions are avoided so that the position of the equilibrium is not too far in favour of the protonated amine and allows a reasonable equilibrium concentration of the free amine, which under most conditions is the reactive species as discussion of the reaction mechanism will demonstrate. There is therefore an optimum level of acidity for the diazotisation of a particular aromatic amine, which depends on the basicity of the amine in question. In the case of aniline derivatives, electron-withdrawing groups, such as the nitro group, reduce the basicity of the amino group. As a consequence, for example, the diazotisation of 4-nitroaniline requires much more acidic conditions than aniline itself. Very weakly basic amines, such as 2,4-dinitroaniline, require extremely acidic conditions. They are usually diazotised using a solution of sodium nitrite in concentrated sulfuric acid, which forms nitrosyl sulfuric acid ($NO^+HSO_4^-$). In recent years, heterocyclic aromatic amines, such as aminothiophenes, aminothiazoles and aminobenzothiazoles, have assumed much greater importance as diazo components, particularly in the synthesis of azo disperse dyes. Diazotisation of these amines can prove more problematic. Generally, the use of concentrated acids is required due to the reduction in basicity of the amine by the heterocyclic system and as a result of protonation of heterocyclic nitrogen atoms, and also because the diazonium salts are sensitive to hydrolysis in dilute acids.

It is especially important to control the acidity when aromatic diamines are treated with nitrous acid to form either the *mono* or *bis* diazonium salts, a process of some importance in the synthesis of disazo dyes and pigments (see later). *p*-Phenylenediamine is an example of a

diamine in which either one or both of the amino groups may be diazotised by careful selection of reaction conditions. The use of dilute
hydrochloric acid can result in smooth formation of the monodiazonium
salt. The use of nitrosyl sulfuric acid is required to diazotise the second
amino group, since the strong electron-withdrawing effect of the diazo
group in the monodiazonium salt reduces the basicity of the amino group
that remains.

It is critically important in diazotisation reactions to maintain careful
control of the temperature of the reaction medium. The reactions are
generally carried out in the temperature range 0–5 °C, necessitating the
use of ice-cooling. In some cases, for example with some heterocyclic
amines, even lower temperatures are desirable, although temperatures
which are too low can cause the reactions to become impracticably slow.
Efficient cooling is therefore essential, not least because the reactions are
invariably strongly exothermic. One reason for the need for low temperatures is that higher temperatures promote the decomposition of nitrous
acid, giving rise to the formation of oxides of nitrogen. The main reason,
however, is the instability of diazonium salts. The diazonium cation,
although stabilised by resonance, decomposes readily with the evolution
of nitrogen, the principal decomposition product being the phenol as
illustrated in Scheme 3.3.

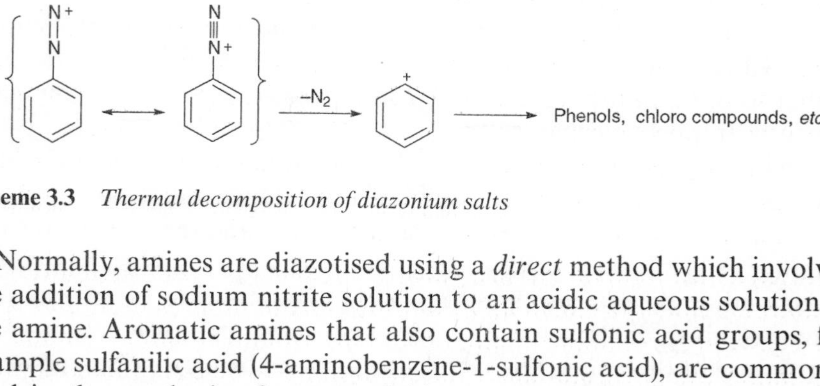

Scheme 3.3   *Thermal decomposition of diazonium salts*

Normally, amines are diazotised using a *direct* method which involves
the addition of sodium nitrite solution to an acidic aqueous solution of
the amine. Aromatic amines that also contain sulfonic acid groups, for
example sulfanilic acid (4-aminobenzene-1-sulfonic acid), are commonly
used in the synthesis of water-soluble azo dyes and in metal salt azo
pigments. Because these amines often dissolve with difficulty in aqueous
acid, they are commonly diazotised using an *indirect* method, which
involves dissolving the compound in aqueous alkali as the sodium salt of
the sulfonic acid, adding the appropriate quantity of sodium nitrite and
adding this combined solution with stirring to the dilute acid.

The quantity of sodium nitrite used in diazotisation is usually the
equimolar amount required by reaction stoichiometry or in very slight
excess. A large excess of nitrite is avoided because of the instability of

nitrous acid and since high concentrations can promote diazonium salt decomposition. When direct diazotisation is used, the sodium nitrite is usually added at a controlled rate such that slight excess is maintained throughout the reaction. In practice, this is monitored easily by the characteristic blue colour which nitrous acid gives with starch/potassium iodide paper. When diazotisation is judged to be complete, any remaining nitrous acid excess is destroyed before azo coupling to avoid side-products due to C-nitrosation of the coupling components. This is usually achieved by addition of sulfamic acid, which reacts as follows:

$$NH_2SO_3H + HNO_2 \rightarrow N_2 + H_2SO_4 + H_2O$$

Because diazonium salts are relatively unstable species, they are almost always prepared in solution as required and used immediately to synthesise an azo dye or pigment. It is generally inadvisable to attempt to isolate diazonium chlorides as they may decompose explosively in the solid state. It is, however, possible to prepare stabilised diazonium salts, which may be handled reasonably safely in the solid state. This is achieved by the use of alternative counter-anions, which are much larger in size and less nucleophilic than the chloride anion. The most commonly used stabilised diazonium salts are tetrafluoroborates (BF_4^-), tetrachlorozincates ($ZnCl_4^{2-}$) and salts obtained from the di-anion of naphthalene-1,5-disulfonic acid. One use of stabilised diazonium salts is in the azoic dyeing of cotton. This process involves the impregnation of cotton fibres with a solution of a coupling component and subsequent treatment with a solution of the stabilised diazonium salt to form an azo pigment, which is trapped mechanically within the cotton fibres. Azoic dyeing, which many years ago was an important method for producing washfast dyeings on cotton, is of relatively limited importance today having largely been superseded by processes such as vat dyeing and reactive dyeing (see Chapter 7).

An outline general mechanism for the diazotisation of an aromatic amine is given in Scheme 3.4. The first stage in the reaction is N-nitrosation of the amine, the nitrosating species being represented in the scheme as Y–N=O. It has been shown that a variety of species may be responsible for nitrosation, depending on the nature of the aromatic amine in question and on the conditions employed for the reaction. Commonly, the nitrosating species may be the nitrosoacidium ion (H_2O^+–NO), nitrosyl chloride (NOCl), dinitrogen trioxide (N_2O_3) or the nitrosonium cation (NO^+). The formation of each of these species from nitrous acid is illustrated in the series of equilibria shown in Scheme 3.5.

The diazotisation reaction provides a classical example of the application of physical chemistry in the elucidation of the detail of organic

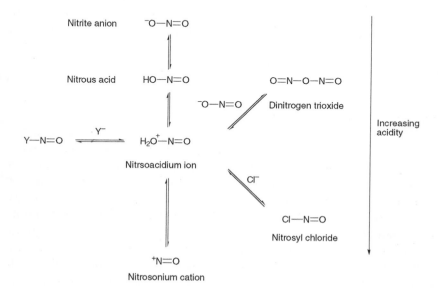

Scheme 3.4 *Mechanism of diazotisation*

Scheme 3.5 *Equilibria involved in the formation of nitrosating species from nitrous acid*

reaction mechanisms. In particular, the results of studies of the kinetics of diazotisation have proved especially informative in establishing the nature of the nitrosating species and the rate-determining step for particular cases. For diazotisation in dilute acids such as sulfuric and perchloric, where the anion is relatively weakly nucleophilic, the nitrosating species has been shown to be dinitrogen trioxide under conditions of low acidity, and the nitrosoacidium ion at higher acidities. In hydrochloric acid, the rate of diazotisation shows a marked increase as a result of catalysis by the chloride anion. The kinetics of the reaction in this case is consistent with a mechanism involving nitrosation of the free amine by reaction with nitrosyl chloride.

In most of the cases discussed so far, the rate-determining step of the reaction is nitrosation of the free amine. When diazotisation is carried out in concentrated acids, the nitrosonium cation, NO^+, is the nitrosating species. In this case, an exchange mechanism has been proposed, as

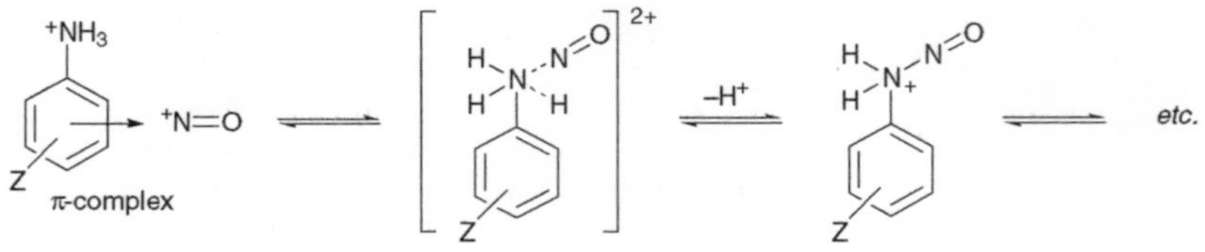

Scheme 3.6 *Mechanism of diazotisation in concentrated acids*

illustrated in Scheme 3.6, in which the initial step is reaction of the nitrosonium cation with the protonated amine to form a π-complex. The deprotonation step which follows the exchange becomes rate determining.

Azo Coupling

Azo coupling is an example of aromatic electrophilic substitution in which the electrophile is the diazonium cation, ArN_2^+. Electrophilic substitution reactions, of which nitration, sulfonation and halogenation are arguably the best-known examples, are the most commonly encountered group of reactions of aromatic systems. However, the diazonium cation is a relatively weak electrophile and will therefore only react with aromatic systems which are highly activated to electrophilic attack by the presence of strongly electron-releasing, groups. The most commonly encountered strongly electron-releasing groups are the hydroxy and amino groups, and this in turn means that the most common compounds which are capable of undergoing azo coupling, referred to as *coupling components*, are either phenols or aromatic amines (either primary, secondary or tertiary). There is a third type of coupling component, commonly a β-ketoacid derivative, in which coupling takes place at a reactive methylene group. The azo coupling reaction between benzenediazonium chloride and phenol is illustrated in Scheme 3.7.

To ensure that the azo dyes and pigments are obtained in high yield and purity, careful control of experimental conditions is essential to minimise the formation of side-products. It is a useful feature of both diazotisation and azo coupling reactions that they may be carried out in water as the reaction solvent. Temperature control, which is so critical in diazotisation reactions, is generally less important in the case of azo coupling. The reactions are normally carried out at or just below ambient temperatures. There is usually little advantage in raising the temperature, other than in a few special cases, since this tends to increase the rate of diazonium salt decomposition more than the rate of coupling.

The experimental factor that requires most careful control in azo

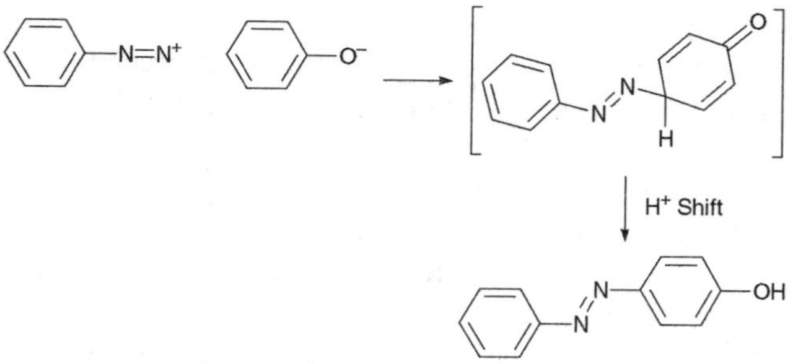

Scheme 3.7   *The azo coupling reaction between benzenediazonium chloride and phenol*

coupling is pH. There is usually an optimum pH range for a specific azo coupling reaction, which is principally dependent on the particular coupling component used. Phenols are usually coupled under alkaline conditions, in which case the phenol (ArOH) is converted predominantly into the phenolate anion (Ar–O⁻). There are two reasons why this facilitates the reaction. The first is a practical reason in that the anionic species is more water-soluble than the phenol itself. A second and arguably more important reason is that the –O⁻ group is more powerfully electron-releasing than the –OH group itself and hence much more strongly activates the system towards electrophilic substitution. Very highly alkaline conditions must be avoided, however, as they promote diazonium salt decomposition. In addition, this can cause conversion of the diazonium cation (ArN_2^+) into the diazotate anion (Ar–N=N–O⁻), a species which is significantly less reactive than the diazonium cation in azo coupling. Generally, it is desirable to carry out the reaction at the lowest pH at which coupling takes place at a reasonable rate. In the case of aromatic amines as coupling components, weakly acidic to neutral conditions are commonly used. The pH is selected such that the amine is converted substantially into the more water-soluble protonated form $(ArNH_3^+)$, but at which there is a significant equilibrium concentration of the free amine $(ArNH_2)$ which is more reactive to towards azo coupling. Reactive methylene-based coupling components undergo azo coupling *via* the enolate anion, the concentration of which increases with increasing pH. These compounds are also frequently coupled at weakly acidic to neutral pH values, under which conditions a sufficiently high concentration of enolate anion exists for the reaction to proceed at a reasonable rate and side-reactions due to diazonium salt decomposition are minimised. Commonly, the rate of addition of the diazonium salt solution to the coupling component is controlled carefully to ensure that an excess of diazonium salt is never allowed to build up in the coupling medium, to minimise side-reactions due to diazonium salt decomposition especially when higher pH conditions are required. This is especially important in the synthesis of azo pigments, insoluble compounds from which the removal of impurities is difficult.

Figure 3.4 illustrates the structures of a range of coupling components commonly used in the synthesis of azo dyes and pigments. In the figure, the position(s) at which azo coupling normally takes place is also indicated. The coupling position is governed by the normal substituent directing effects, both electronic and steric, encountered for aromatic electrophilic substitution. These effects, together with other aspects of the reaction mechanisms involved in electrophilic aromatic substitution are dealt with at length in most organic chemistry textbooks, and so are not considered further here. The coupling components include the relatively simple benzene derivatives, phenol (**30**) and aniline (**31**), naphthalene derivatives **32–35**, some heterocyclic compounds such as the pyrazolones **36** and pyridones **37**, while the β-keto acid derivatives are exemplified by acetoacetanilide (**38**). Many coupling components, such as compounds **30–32** and **36–38**, are capable of a single azo coupling reaction to give a monoazo colorant. A number of coupling components, for example naphthalene derivatives **33–35**, contain an amino and a hydroxy group in separate rings. These compounds are useful because they are capable of

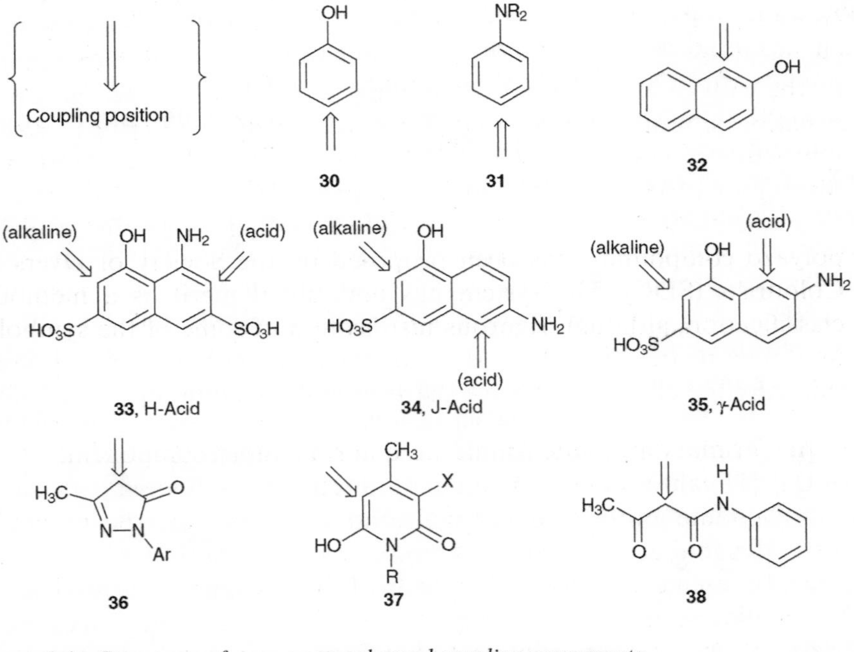

30 **31** **32**

33, H-Acid **34**, J-Acid **35**, γ-Acid

36 **37** **38**

Figure 3.4 *Structures of some commonly used coupling components*

reacting twice with diazonium salts, thus providing a route to disazo colorants. As an example, 1-amino-8-hydroxynaphthalene-3,6-disulfonic acid, **33**, referred to trivially as H-Acid, is used in the synthesis of a number of important azo dyes. The position of azo coupling with this coupling component may be controlled by careful choice of pH. Under alkaline conditions, the hydroxy group is converted into the phenolate anion ($-O^-$) which is more electron releasing than the amino group. In contrast, under weakly acidic conditions it exists un-ionised as the –OH group which is less electron releasing than the amino group. A feature of naphthalene chemistry is that a substituent exerts its maximum electronic effect in the ring to which it is attached, so that under alkaline conditions, azo coupling is directed into the ring containing the hydroxy group, while under weakly acidic conditions reaction takes place in the ring containing the amino group. This selectivity of the coupling position, also shown by coupling components such as J-Acid (**34**) and γ-Acid (**35**), allows the preparation in a controlled manner of a range of unsymmetrical disazo dyes.

STRATEGIES FOR AZO DYE AND PIGMENT SYNTHESIS

In the case of monoazo dyes and pigments, the strategy for synthesis is straightforward, involving appropriate selection of diazo and coupling components, and choice of reaction conditions in accordance with the chemical principles presented in the previous two sections of this chapter. In the case of azo colorants containing more than one azo group, the situation is more complex and it becomes even more critical that the synthetic strategy and reaction conditions are selected carefully to ensure that a pure product is obtained in high yield. A system, in common usage as a systematic treatment of the possible synthetic strategies leading to polyazo compounds, has been proposed by the Society of Dyers and Colourists (SDC). The system has undoubted merit as a method of classification, although rigorous justification of some of the symbolism used may be questioned. To describe the strategies, it is appropriate at this stage to define of the nature of the various reacting species as follows:

A: primary aromatic **A**mine: *i.e.* a normal diazo component;
D: primary aromatic **D**iamine: *i.e.* a tetrazo component;
E: coupling component capable of reaction with one diazonium ion: an **E**nd component;
Z: coupling component capable of reaction with more than one diazonium ion;
M: coupling component containing a primary aromatic amino group

which, after an azo coupling reaction, may be diazotised and used in a second azo coupling: a Middle component.

Synthesis of Monoazo Dyes and Pigments

The synthesis of monoazo dyes and pigments is represented by the symbolism

$$A \rightarrow E$$

In using this terminology, it should be emphasised that the arrow has the meaning 'diazotised and then coupled with' rather than its usual meaning in organic reaction sequences. Thus, for example, the first stage in the synthesis of C. I. Disperse Orange 25, **39**, is the diazotisation of 4–nitroaniline, using sodium nitrite and aqueous hydrochloric acid at temperatures less than 5 °C. The diazonium salt thus formed is reacted with N-ethyl-N-β-cyanoethylaniline under weakly acidic conditions since the coupling component is a tertiary aromatic amine.

39

Synthesis of Disazo Dyes and Pigments

The situation becomes more complex when two separate diazotisations and azo couplings are required. Four separate strategies leading to disazo colorants may be identified, using the commonly accepted SDC terminology and symbolism, exemplified as follows as strategies (a)–(d).

(a) $A^1 \rightarrow Z \leftarrow A^2$

In this the first strategy, two primary aromatic amines (A^1 and A^2) are diazotised and reacted separately under appropriate pH conditions and in an appropriate sequence with a coupling component which has two available coupling positions. As an example, the synthesis of C. I. Acid Black 1, **40**, a bluish-black dye commonly used to dye wool, is as follows. 4-Nitroaniline is diazotised under the usual conditions and the diazonium salt reacted in the first coupling reaction under weakly acidic conditions with H-Acid, **33**, under which conditions azo coupling is directed into the ring containing the amino group. Secondly, aniline is diazotised and the resulting diazonium salt reacted with the monoazo intermediate to form the disazo dye. For the second coupling, alkaline

40

conditions are used since a phenolic coupling component is involved. In general, reactions of this type are carried out in this sequence because of the good solubility of the monoazo intermediate in alkali and the improved selectivity of the process when carried out in this way.

(b) $E^1 \leftarrow D \rightarrow E^2$

In this strategy a primary aromatic diamine is diazotised twice (tetrazotised) and coupled separately with two coupling components (E^1 and E^2). In the case of C. I. Pigment Yellow 12, **41**, which is an important bright yellow pigment used extensively in printing inks, the product is symmetrical (a bisketohydrazone). This greatly simplifies the synthetic procedure since the two coupling reactions may be carried out simultaneously. 3,3'-Dichlorobenzidine (1 mol) is tetrazotised (bisdiazotised) and coupled under weakly acidic to neutral conditions with acetoacetanilide (**38**, 2 mol) to give the product directly. When an unsymmetrical product is required, a much more careful approach to the synthesis is essential to ensure that the unsymmetrical product is not contaminated by quantities of the two possible symmetrical products. One approach that may be used involves the diazotisation, by careful choice of conditions, of one amino group of the diamine D followed by coupling with the first component E^1. The monoazo intermediate formed contains the second amino group, which may be diazotised followed by coupling with component E^2. Alternatively, the synthesis may be achieved by tetrazotisation of the diamine followed by two sequential azo coupling reactions with careful selection of pH conditions to control the outcome of the reactions. In the case of C. I. Direct Blue 2, **42**, for example, the synthesis could start, in principle, with the tetrazotisation of benzidine (4,4'-diaminobiphenyl). The resulting tetrazonium salt is first coupled with γ-Acid (**35**) under acidic conditions and then the monoazo intermediate is reacted with H-Acid (**33**) under alkaline conditions. This particular example, by way of illustration, uses benzidine as the tetrazo component. Formerly, benzidine was an important tetrazo component, particularly for the manufacture of direct dyes. However, the compound has for many years now been recognised as a potent human carcinogen and its use in colour

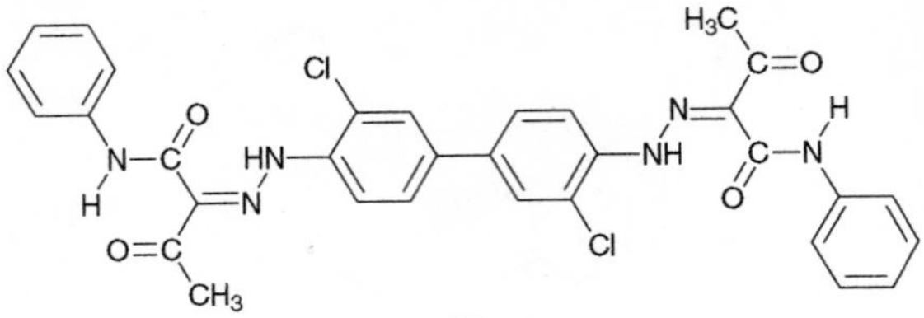

41

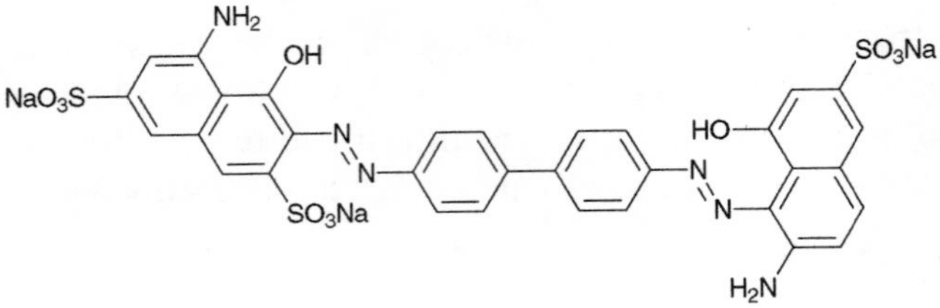

42

manufacture has long since ceased in the developed world (see Chapter 11).

(c) A → M → E

This third strategy in azo colorant synthesis makes use of the feature that a primary aromatic amine has the potential to be used both as a coupling component and as a diazo component. As an example, in the synthesis of C. I. Disperse Yellow 23, **43**, aniline is first diazotised and the resulting diazonium salt reacted under weakly acidic conditions with aniline to give 4-aminoazobenzene. A practical complication with this stage is the inefficiency of the coupling reaction with aniline, and the formation of side-products due to an *N*-coupling reaction. In an improved method, aniline is first reacted with formaldehyde and sodium bisulfite to form the methyl-ω-sulfonate derivative ($ArNHCH_2SO_3Na$) which couples readily. After coupling, the labile methyl-ω-sulfonate group may be removed easily by acid hydrolysis to give 4-aminoazobenzene. This amine is then diazotised and the resulting diazonium salt is reacted with phenol under alkaline conditions to give the disazo compound **43**.

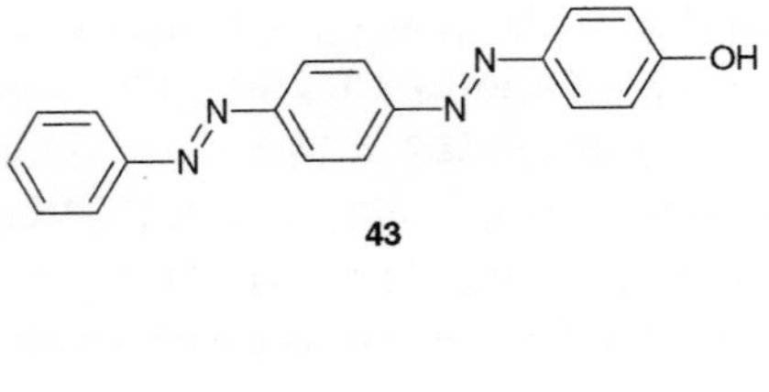

43

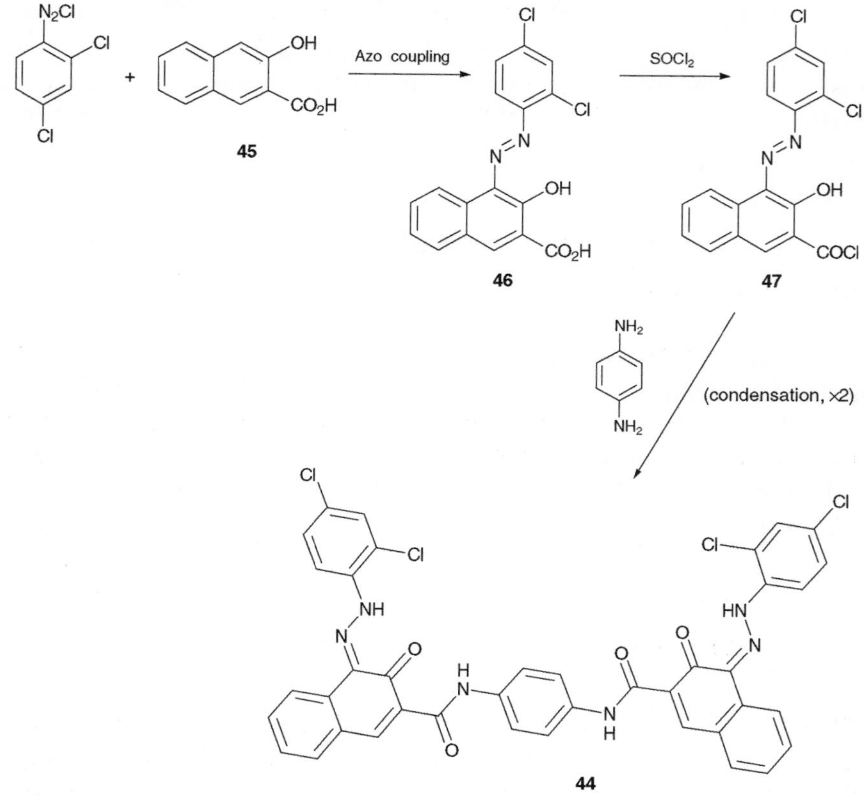

Scheme 3.8 *Synthesis of a disazo condensation pigment*

(d)   A¹ → Z–X–Z ← A²

The products from the fourth strategy are structurally not dissimilar from those obtained by the strategy given in (a). However, in this strategy the disazo colorant is synthesised by linking together two molecules of a monoazo derivative by some chemical means. As an example, the synthesis of C. I. Pigment Red 166, **44**, a disazo condensation pigment which exists structurally as a bisketohydrazone, is shown in Scheme 3.8. The monoazo compound **46** containing a carboxylic acid group is prepared by an azo coupling reaction with 3-hydroxy-2-naphthoic acid (**45**) as the coupling component. The acid **46** is then converted into the acid chloride **47**, followed by a condensation reaction between the acid chloride (2 mol) and *p*-phenylenediamine (1 mol) to form compound **44**. In principle, a simpler and more cost-effective route using strategy (a), involving azo coupling of the diazonium salt (2 mol) with the appropriate bifunctional coupling component (1 mol), might be proposed. In practice, this route fails because the monoazo derivative formed from the first azo coupling reaction is so insoluble that the second coupling reaction cannot take place.

Dyes and Pigments Containing More than Two Azo Groups

The strategies leading to dyes with several azo groups are in essence extensions of the methods leading to disazo colorants. As the number of separate diazotisations and azo coupling reactions required for the synthesis of a polyazo dye increases, so does the number of potential strategies. There are a number of commercial trisazo dyes, most being brown or black direct dyes for application to cotton. There are five strategies leading to trisazo colorants, which may be illustrated using the accepted symbolism as:

(a) $E \leftarrow D \rightarrow Z \leftarrow A$
(b) $E^1 \leftarrow D \rightarrow M \rightarrow E^2$
(c) $A \rightarrow M^1 \rightarrow M^2 \rightarrow E$
(d) $A^1 \rightarrow M \rightarrow Z \leftarrow A^2$
(e) $A^1 \rightarrow Z \leftarrow A^2$
$$\uparrow$$
$$A^3$$

As an example of strategy (a), the synthesis of C. I. Direct Black 38, **48**, may be represented as follows:

m-phenylenediamine $\leftarrow$ benzidine $\rightarrow$ H-Acid $\leftarrow$ aniline

A few azo dyes containing four, five and six azo groups, referred to respectively as tetrakisazo, pentakisazo and hexakisazo dyes, are known although they are progressively less important commercially and, naturally, the synthetic strategies are more complex. Also, because of the

48

increased likelihood of side-reactions, the products are increasingly less likely to be pure species.

METAL COMPLEX AZO DYES AND PIGMENTS

Metal complex formation has been a prominent feature of textile dyeing from very early times, since it was recognised that the technical perform-ance, including fastness to washing and light, of many natural dyes could be enhanced by treatment with certain metal ions, a process known as *mordanting*. Mordant dyeing is still used to a certain extent today, al-though it is restricted mainly to the complexing of certain azo dyes on wool with chromium(III) (see Chapter 7). There are basically two types of metal complex azo dyes: those in which the azo group is coordinated to the metal (medially metallised) and those in which it is not (terminally metallised). The discussion in this section is restricted to the former type which are by far the most important commercially.

The most important metal complex azo dyes are formed from the reaction of transition metal ions with ligands in which the *ortho* positions adjacent to the azo group contain groups capable of coordinating with the metal ion. The most important group in this respect is the hydroxy (–OH) group, although carboxy ($-CO_2-$) and amino (–NH–) groups can also be used. The most important transition metals used commercially to form metal complex azo dyes are copper(II), cobalt(III) and, especially, chromium(III). The reaction of an *o,o'*-dihydroxyazo compound, which acts as a tridentate ligand, with chromium(III) is illustrated in Figure 3.5. The azo group is capable of coordination to the metal ion through only one of the two azo nitrogen atoms, utilising its lone pair in bonding. While transition metal complex azo chemistry has proved to be of con-siderable importance in textile dyes, curiously, and perhaps rather sur-prisingly, it has provided very limited success in commercial organic pigments, although other types of metal complexes, notably phthalocyanines, are of immense significance in this respect (see Chapters 5 and 9).

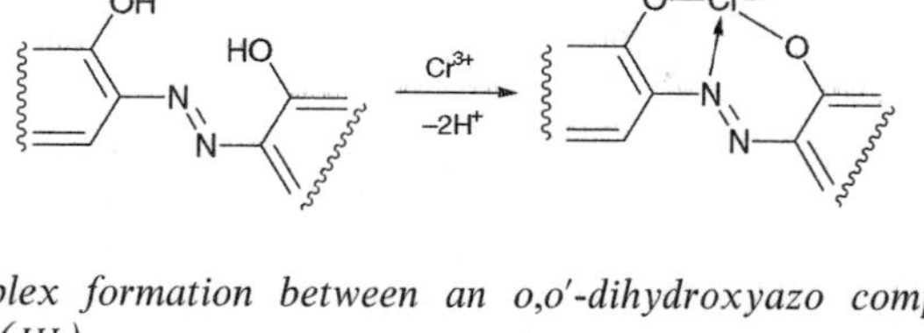

Figure 3.5 *Complex formation between an o,o'-dihydroxyazo compound and chro-mium(III)*

1:1 Copper complexes of azo dyes are used widely in both reactive dyes (Chapter 8) and direct dyes (Chapter 7) for cotton. In these dyes, the copper complexes adopt a four-coordinate square planar geometry, with the three coordinating sites of the dye occupying three corners of the square and the fourth occupied by a monodentate ligand, commonly water (Figure 3.6). The most important cobalt and chromium complexes of azo dyes adopt six-coordinate octahedral geometry, with the six positions occupied by coordination with two tridentate azo dye ligands. These products are referred to as 2:1 premetallised dyes and they are of considerable importance in dyeing protein fibres such as wool (Chapter 7).

The 2:1 octahedral complexes present many opportunities for isomerism. The isomerism in a number of metal complex azo dyes has been characterised using a variety of techniques, including ^{1}H and ^{15}N NMR spectroscopy and X-ray crystallography, and in some cases isomers have been separated chromatographically. The geometrical isomerism of some metal complex azo dyes is illustrated schematically in Figure 3.7. There are a number of geometrical isomers possible: one meridial (*mer*) isomer (I), in which the two dye molecules are mutually perpendicular, and five facial (*fac*) (II–VI) isomers in which the molecules are parallel. The meridial arrangement is favoured by complexes of *o,o'*-dihydroxyazo systems that contain a 5:6 chelate ring system, while the facial arrangement is favoured in *o*-carboxy-*o'*-hydroxyazo systems in which there is a 6:6 chelate ring system (Figure 3.8). Because of their asymmetry, the geometrical isomers, apart from (VI) which is centrosymmetric, also give rise to pairs of optically-active enantiomers. Other types of isomerism in azo metal complexes which have been identified include positional isomerism as a result of the possibility of coordination to either (but not both) of the azo nitrogen atoms, and isomerism as a result of different states of hybridisation (sp^2 or sp^3) of the azo nitrogen atoms.

There are a number of particular technical advantages associated with the formation of coloured metal complexes. Commonly, the transition metal complexes of a coloured organic ligand exhibit lightfastness which is significantly better than that of the free ligand. An explanation that has been offered for this effect is that coordination with a transition metal ion

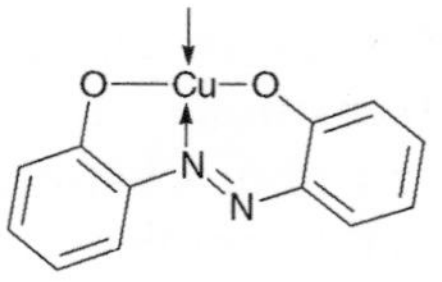

Figure 3.6   *Square planar azo copper complex*

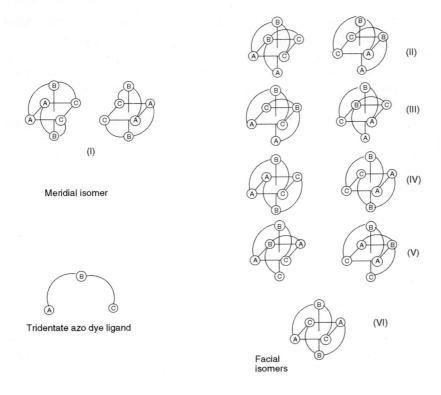

Figure 3.7 *Isomerism in 2:1 octahedral metal complex azo dyes*

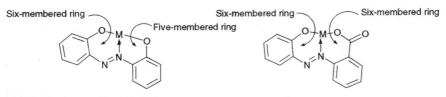

Figure 3.8 *Chelate ring formation in metal complex azo dyes*

reduces the electron density at the chromophore, which in turn leads to improved resistance to photochemical oxidation. Other effects which may have a part to play in the enhanced lightfastness of transition metal complexes are steric protection of the chromophore towards degrading influences and the ability of transition metal ions to quench excited states that otherwise might undergo photochemical decomposition. In addition, the larger size of the metal complex molecules compared with the free ligand generally gives rise to improved washfastness properties in textile dyes as a result of stronger interactions with the fibre. On the other hand, the colours of the metal complexes are almost invariably duller than those of the azo dye ligand, a feature that limits their usefulness. The reduction the brightness of the colour which accompanies metal complex formation is due to a broadening of the visible absorption band. There are a number of possible reasons for this effect. Broadening may due to the presence of a number of isomers, each with a slightly different absorption band. Alternatively, it may be due to overlap of the absorption band associated with the π–π^* transitions of the ligand with those arising from metal ion d–d transitions or from ligand–metal charge transfer transitions.

Chapter 4

Carbonyl Dyes and Pigments

The chemical class of colorants which is second in importance to azo dyes and pigments is characterised by the presence of a carbonyl (C=O) group, which may be regarded as the essential chromophoric unit. The vast majority of carbonyl dyes and pigments contain two or more carbonyl groups which, as illustrated in Figure 4.1, are linked to one another through a conjugated system, frequently an aromatic ring system.

Carbonyl colorants are found in a much wider diversity of structural arrangements than is the case with azo dyes and pigments. By far the most important group of carbonyl dyes and pigments are the anthraquinones, to which a substantial part of this chapter is devoted. Other types which are of commercial importance in particular application classes include indigoids, benzodifuranones, coumarins, naphthalimides, quinacridones, perylenes, perinones and diketopyrrolopyrroles. Carbonyl dyes and pigments are capable of providing a wide range of colours, essentially covering the entire visible spectrum. However, a major reason for the importance of carbonyl colorants is that they are capable of giving long wavelength absorption bands with relatively short conjugated systems. This feature applies especially to anthraquinone and indigo derivatives, which are thus of particular importance in the blue shade area.

In terms of fastness properties, carbonyl dyes and pigments are often superior to their azo counterparts. They are thus commonly the colorants of choice when high technical performance is demanded by a particular

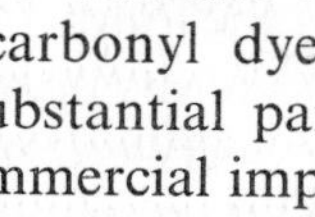

Figure 4.1 *Structural arrangement in most carbonyl colorants*

69

application. In most textile dye application classes, carbonyl dyes (mostly anthraquinones) rank second in significance to azo dyes. A particular textile application class dominated by carbonyl dyes is the vat dye class, a group of dyes applied to cellulosic fibres such as cotton (see Chapter 7). In the vat dyeing of cotton, the ability of the carbonyl groups to undergo a reversible reduction to a water-soluble form is utilised. Azo dyes are inappropriate as vat dyes because of the instability of the azo group to the reduction process. A range of carbonyl pigments, including the quinacridones, perylenes, perinones and diketopyrrolopyrroles, impart extremely high performance in their application. These products owe their high fastness to light, heat and solvents to the ability of the carbonyl group to participate in intra- and intermolecular hydrogen bonding.

The chemistry involved in the manufacture of carbonyl colorants is generally more elaborate and much less versatile than is the case with azo dyes. Often the synthetic sequence involves multiple stages and the use of specialist intermediates. Consequently, the number of commercial products is more restricted and they tend to be rather more expensive. Indeed, certain carbonyl dyes, notably some anthraquinones, are becoming progressively less important commercially, particularly for traditional textile applications, as a wider range of azo dyes absorbing at longer wavelengths have emerged and as the cost differential between azo dyes and carbonyl dyes increases. In the following sections of this chapter, the characteristic structural features of the most important types of carbonyl colorants are reviewed and an overview of some of the more important synthetic routes is presented.

ANTHRAQUINONES

A common arrangement of the carbonyl groups in coloured molecules gives rise to a group of compounds known as *quinones*. These may be defined as cyclohexadienediones, *i.e.* compounds containing two ketone carbonyl groups and two double bonds in a six-membered ring. The simplest quinones are *o*- and *p*-benzoquinones, **49** and **50** respectively. Derivatives of the benzoquinones and of the naphtho-1,4-quinone system **51** are of only minor interest as colouring materials. This is probably due, at least in part, to the instability which results from the presence of an alkene-type double bond. By far the most important quinone colorants are the anthraquinones, or more correctly anthra-9,10-quinones. Anthraquinones, as demonstrated by the parent compound **52**, contain a characteristic system of three linear fused six-membered rings in which the carbonyl groups are in the central ring and the two outer rings are fully aromatic. Anthraquinone colorants can give rise to the complete range of hues. However, within any particular application class, they are often more important as violets, blues and greens, thus complementing the azo chemical class which generally provides the most important yellows, oranges and reds. They commonly give brighter colours than azo dyes, but the colours are often weaker, one reason why they are in general less cost-effective than azo dyes. Anthraquinone dyes provide excellent lightfastness properties, generally superior to azo dyes.

There is a wide diversity of chemical structures of anthraquinone colorants. Many anthraquinone dyes are found in nature, perhaps the best known being alizarin, 1,2-dihydroxyanthraquinone, the principal constituent of madder (see Chapter 1). These natural anthraquinone dyes are no longer of significant commercial importance. Many of the current commercial range of synthetic anthraquinone dyes are simply substituted derivatives of the anthraquinone system. For example, a number of the most important red and blue disperse dyes for application to polyester fibres are simple non-ionic anthraquinone molecules, containing substituents such as amino, hydroxy and methoxy, and a number of sulfonated derivatives are commonly used as acid dyes for wool.

There are a large number of anthraquinones which are structurally more complex and polycyclic in nature. In this book, provided that the anthraquinone nucleus is recognisable somewhere in the structure, the colorant is classed as a member of the anthraquinone class, although some texts consider these annellated derivatives separately. These polycyclic anthraquinones belong almost invariably to the vat dye textile application class where their large extended planar structure is an important feature for their application (see Chapter 7). There are around 200 different anthraquinonoid vat dyes in use commercially, of widely different structural types, and covering the entire spectrum from yellow through blue to black. Figure 4.2 shows a selection of these compounds, including both carbocyclic and heterocyclic types. The examples presented are indanthrone (**53**), C. I. Vat Blue 4, violanthrone (**54**), C. I. Vat Blue 20, pyranthrone (**55a**), C. I. Vat Orange 9, flavanthrone (**55b**), C. I. Vat Yellow 1 and C. I. Vat Green 8, (**56**), a derivative containing 19 fused rings. While the anthraquinone chromophoric grouping is second only in importance to the azo chromophore in the chemistry of textile dyes, it is of considerably less importance in pigments. This is probably due to the

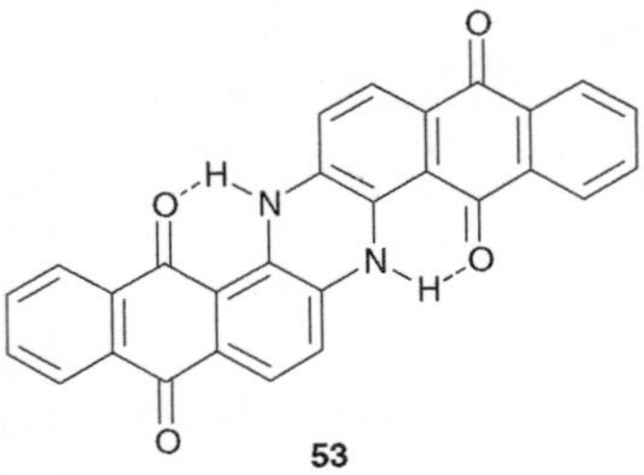

Figure 4.2 *Some polycyclic anthraquinone vat dyes and pigments*

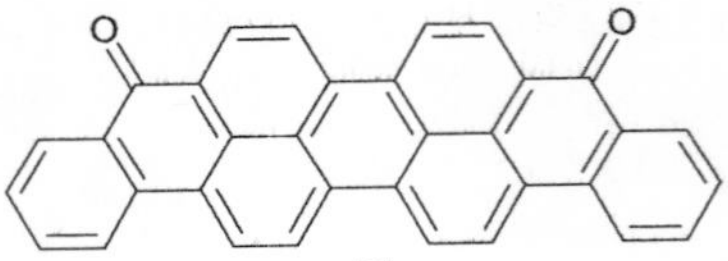

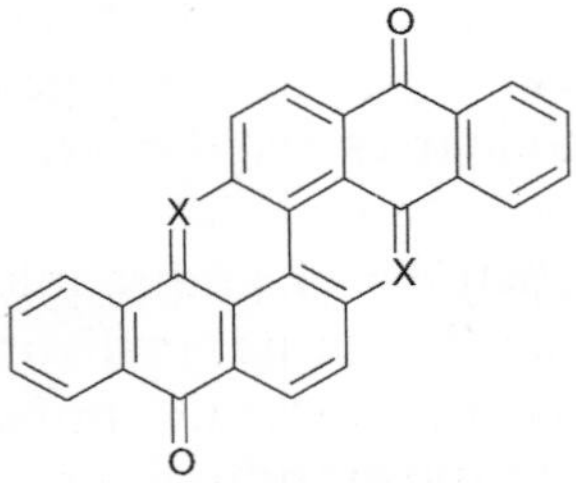

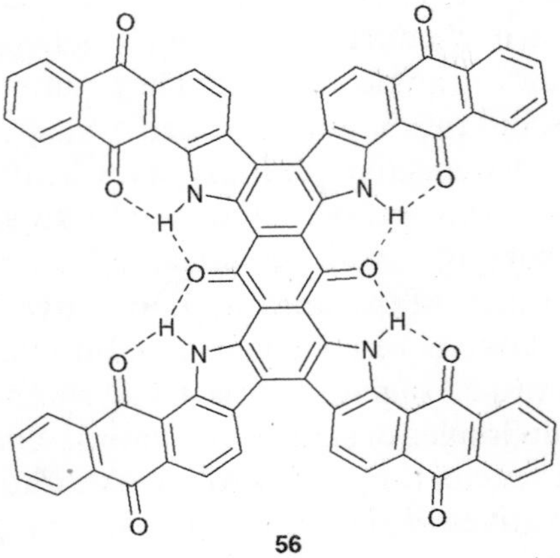

fact that the traditional role of anthraquinones in many dye application classes, which is to provide lightfast blues and greens, is more successfully adopted by the phthalocyanines (Chapter 5) in the case of pigments. However, the insolubility and generally good fastness properties of vat dyes stimulated considerable effort into the selection of suitable examples of the colorants for use, after conversion into the appropriate physical form, as pigments for paint, printing ink and plastics applications. Of the known vat dyes, only about 25 have been fully converted into pigment use, and less than this number are of real commercial significance. The range of anthraquinone pigments includes some of the longest-established vat dyes, notably indanthrone (**53**, C. I. Pigment Blue 60) together with some of its halogenated derivatives, and flavanthrone (**55b**, C. I. Pigment Yellow 24).

Anthraquinone (**52**) is only weakly coloured, its strongest absorption being in the UV region (λ_{max} 325 nm). The UV/visible spectral data for a series of substituted anthraquinones, **52a–h**, are given in Table 4.1 and these illustrate the effect of the substituent pattern on the colour. The introduction of simple electron-releasing groups, commonly amino or

Table 4.1 *Absorption maxima for some substituted anthraquinones*

Compound	Substituent	λ_{max} (nm) in MeOH
52a	1-OH	402
52b	2-OH	368
52c	1-NH$_2$	475
52d	2-NH$_2$	440
52e	1-NHPh	500
52f	1,4-diNH$_2$	590
52g	1,4,5,8-tetraNH$_2$	610
52h	1,4-diNHPh	620

hydroxy, into the anthraquinone nucleus gives rise to a bathochromic shift which is dependent on the number and position of the electron-releasing group(s) and their relative strengths (OH < NH$_2$ < NHAr). They are thus typical donor–acceptor systems, with the carbonyl groups as the acceptors and the electron-releasing auxochromes as the donors. By choice of an appropriate substitution pattern, dyes are obtained which may absorb in any desired region of the visible spectrum. From the data, it is clear that the electron-releasing groups exert their maximum bathochromic effect in the α-positions (1-, 4-, 5-, 8-) rather than the β-positions (2-, 3-, 6-, 7-). This is one reason why substitution in α-positions is preferred in anthraquinone dyes. In addition, α-substituents give dyes with higher molar extinction coefficients and, very importantly, they enhance technical performance, especially lightfastness, by virtue of their participation in intramolecular hydrogen bonding with the carbonyl groups. The 1,4-substitution pattern is particularly significant in providing blue dyes of commercial importance in a variety of textile dye application classes (Chapter 7).

The effect of substituents on colour in substituted anthraquinones may be rationalised using the valence-bond (resonance) approach, in the same way as has been presented previously for a series of azo dyes (see Chapter 2 for details). For the purpose of explaining the colour of the dyes, it is assumed that the ground electronic state of the dye most closely resembles the most stable resonance forms, the normal Kekulé-type structures, and that the first excited state of the dye more closely resembles the less stable, charge-separated forms. Some relevant resonance forms for anthraquinones **52**, **52c**, **52d** and **52f** are illustrated in Figure 4.3. The ground state of the parent compound **52** is assumed to resemble closely structures such as I, while charge-separated forms, such as structure II, are assumed to make a major contribution to the first excited state. Structure II is clearly unstable due to the carbocationic centre. In the case of aminoanthraquinones **52c** and **52d**, donation of the lone pair from the

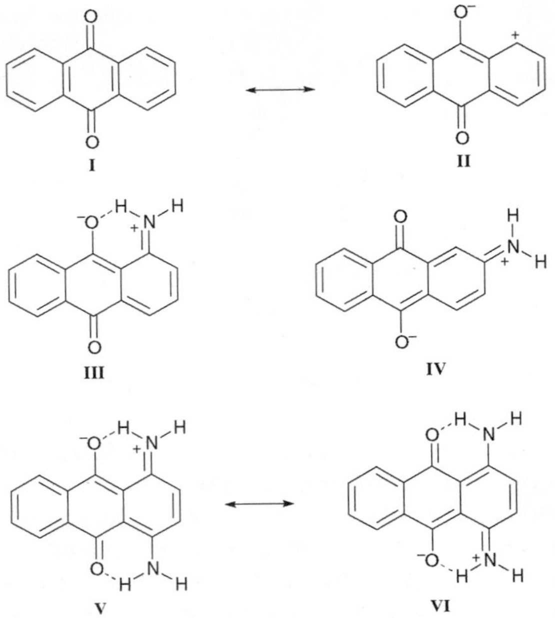

Figure 4.3   *Some relevant resonance forms for anthraquinones* **52, 52c, 52d** *and* **52f**

amino nitrogen atom markedly stabilises the first excited states, represented respectively by structures III and IV, lowering their energy and leading, as a consequence of the inverse relationship between the difference in energy and the absorption wavelength, to a pronounced bathochromic shift. Two reasons may be proposed for the enhanced bathochromicity of the 1-isomer (**52c**), compared with the 2-isomer (**52d**). Firstly the proximity of positive and negative charges in space in the first excited state structure III gives rise to a degree of electrostatic stabilisation. Secondly, the intramolecular hydrogen bonding in structure III, which is not present in structure IV, serves to increase the electron-releasing power of the lone-pair on nitrogen atom and to increase the electron-withdrawing effect of the carbonyl group, both of which effects stabilise the excited state relative to the ground state and lead to a more pronounced bathochromic effect. The 1,4-diamino compound **52f** is particularly bathochromic because of an extensively resonance-stabilised first excited state, involving structures V and VI.

INDIGOID DYES AND PIGMENTS

Indigo (**57**), the parent system of this group of colorants, is one of the oldest known natural dyes (see Chapter 1). The naturally occurring material is indican, the colourless glucoside of indoxyl. When subjected to a fermentation process, free indoxyl (**58**) is generated by enzymic hydrolysis and this compound undergoes oxidation in air to indigo. Indigo was formerly made in this way in India, Indonesia and China from the indigo plant (*Indigofera tinctoria*) and in Europe from dyers' woad (*Isatis tinctoria*) although nowadays, indigo is made purely synthetically. The synthetic routes are outlined later in this chapter.

57   **58**

The structure of indigo was first proposed by von Baeyer, although he initially suggested that it had the (*Z*)- (or *cis*) configuration **59**. An X-ray crystal structure determination carried out by Posner in 1926 confirmed that the molecule in fact exists as the (*E*)- (or *trans*) isomer **57**. Although hundreds of indigoid derivatives have been synthesised and evaluated as dyes over the years, relatively few of these have achieved commercial importance and indeed the parent compound remains by far the most important member of the class. Indigo is applied to fabrics as a vat dye, imparting an attractive blue colour and good wet fastness properties. It provides only moderate lightfastness, but it shows the feature of fading without changing colour (*i.e.* on tone). By far its most important use is to dye jeans and other denim articles, which slowly fade to paler shades of blue, a key fashion feature. In fact, its lasting success as a commercial dye will be heavily dependent on the continuation of this popular textile fashion trend.

59

A question which has intrigued colour chemists for years is why indigo, a relatively small molecule, absorbs at such long wavelengths. The colour of indigo depends crucially on its environment. It is known that, in the vapour phase, the only situation in which it approaches a monomolecular state, indigo is red. In solution, indigo exhibits pronounced positive solvatochromism; in non-polar solvents it is violet, while in polar solvents it is blue. In the solid state, and when applied to fabric as a vat dye, it is blue. X-ray structure analysis of the molecule has demonstrated that, in the solid state, the molecules are highly aggregated by intermolecular hydrogen bonding, each indigo molecule being attached to four others. It is clear that the intermolecular hydrogen bonding is a major factor in providing the bathochromic shift of colour. It also explains its low solubility and relatively high melting point (390–392 °C). It is generally accepted that the basic structural unit responsible for the colour of indigo is as illustrated in Figure 4.4. It consists of two electron donor groups (NH) and two electron acceptor groups (C=O), 'cross-conjugated' through an ethene bridge, the so-called the H-chromophore, and this gives rise to a remarkably bathochromic absorption. The molecule has been the subject of numerous studies using quantum mechanical calculations and these are in agreement with the H-chromophore explanation.

UV/visible spectral data for some indigo derivatives, which illustrate the nature of the substituent effects, are given in Table 4.2. A valence-bond (resonance) explanation has been suggested for the bathochromicity of indigo, invoking important resonance contributions to the first excited state from charge separated structures VII–X, in each of which charge is accommodated in a stable location, as illustrated in Figure 4.5. Resonance structures VII and VIII are particularly stable due to the retention of aromatic character of both benzene rings. The effect of

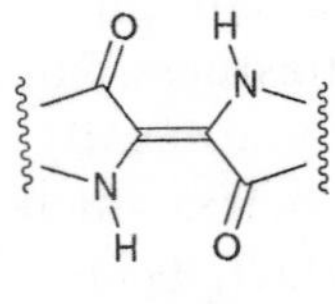

Figure 4.4 *The H-chromophore unit of indigo*

Table 4.2 *UV/visible spectral data (λ_{max} values in nm) for solutions of some symmetrical di-substituted indigo derivatives* **57** *and* **57a–d** *in 1,2-dichloroethane.*

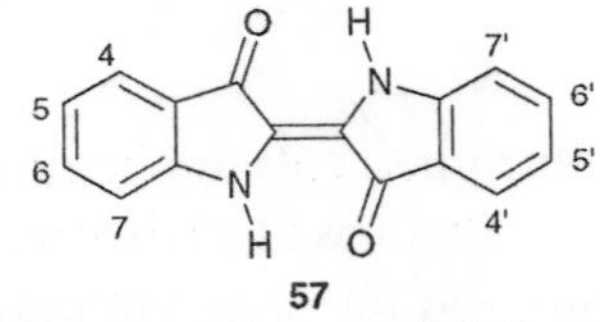

Substituent	5,5'-Isomer	6,6'-Isomer
H	**57**, 620 nm	**57**, 620 nm
NO$_2$	**57a**, 580 nm	**57b**, 635 nm
OCH$_3$	**57c**, 645 nm	**57d**, 570 nm

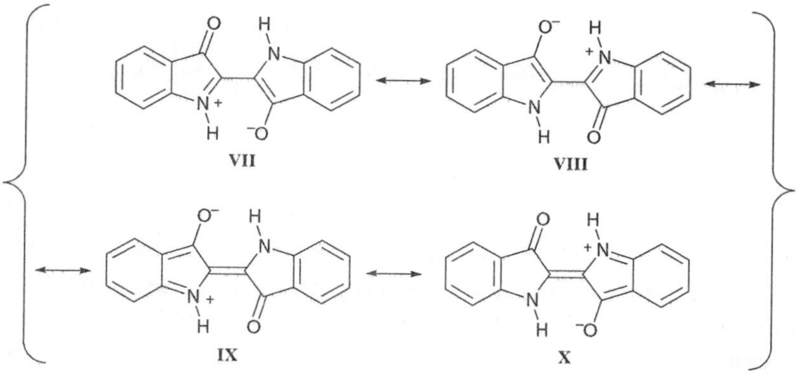

Figure 4.5 *Valence-bond (resonance) approach to indigo*

substituents on the colour of indigo may be explained using this approach. For example, an electron-releasing group in the 5- (or 7)-positions, which are *ortho* or *para* respectively to the NH group stabilise the first excited state by increasing the electron density on the nitrogen atom and a bathochromic shift results. In contrast, an electron-withdrawing group at these positions destabilises the first excited state by increasing the positive charge at an already electron-deficient nitrogen atom, causing a hypsochromic shift. For substituents at the 4- and 6-positions, which are in conjugation with the electron-withdrawing carbonyl group, the situation is reversed. The results of PPP-MO calculations are found, in this case, to support the valence-bond explanation.

A range of other indigoid systems, represented by the general structure **60** in which one or both of the NH groups are replaced by other heteroatoms capable of donating a lone pair of electrons, are known. These include the yellow oxindigo (**60a**, X = Y = O), the red thioindigo (**60b**, X = Y = S), and the violet selenoindigo (**60c**, X = Y = Se), together with mixed derivatives such as compound **60d** (X = NH, Y = S). In this group, thioindigo (**60b**) is a useful red vat dye and some of its halogenated derivatives are used as pigments, but the others are of limited commercial significance. Indirubin (**61**) is a red positional isomer of indigo that is occasionally encountered as an impurity in natural indigo.

BENZODIFURANONES

The benzodifuranones constitute one of the more recently introduced groups of carbonyl dyes. They were launched commercially in the late 1980s by ICI as disperse dyes for application to polyester. This group of dyes, of which compound **62** is a representative example, are capable of

60   61

providing a range of colours but the most important are red. They
provide bright intense red shades on polyester with good fastness to light,
sublimation and washing. For good build-up of colour on polyester, it
has been found that asymmetric products, such as compound **62**, perform
better than when the molecules are symmetrically substituted.

62

FLUORESCENT CARBONYL DYES

A number of carbonyl dyes are characterised by strong fluorescence. One
of the most important groups of fluorescent brightening agents and dyes
is based on the coumarin ring system. The important fluorescent
coumarin derivatives almost invariably contain an electron-releasing
substituent in the 7-position, which is most commonly a dialkylamino
group but may be hydroxy or methoxy. The earliest and still the most
widely-used coumarin dyes contain a benzimidazolyl, **63a**, benzoxazolyl,
63b, or benzothiazolyl, **63c**, group as the acceptor in the 3-position. C. I.
Disperse Yellow 232, **63d**, for example, dyes polyester to give a brilliant
fluorescent greenish-yellow hue with good fastness to light, sublimation
and washing. Other important types of fluorescent carbonyl dyes include
the aminonaphthalimides, such as C. I. Disperse Yellow 11, **64**, and some
sterically hindered perylenes which are considered in the next section.
Fluorescent dyes and pigments find a wide range of uses in textiles,
plastics, paint and printing inks where high visual impact is desirable,
such as advertising and safety applications. They are also used in the

detection of flaws in engineered articles, solar collector systems, dye lasers and a host of analytical and biological applications.

63a X = NH, R = H
63b X = O, R = H
63c X = S, R = H
63d X = O, R = Cl

64

CARBONYL PIGMENTS

There are a number of carbonyl colorant systems, which offer particular advantages as pigments for applications, such as automotive paints, which demand extremely high performance in terms of colouristics and fastness to light, heat and solvents. These carbonyl pigments are diverse structurally and include the quinacridones, diketopyrrolopyrroles, perylenes, perinones and anthraquinones and the first four of these types are discussed in this section. The use of anthraquinones as vat pigments has been discussed in a previous section. Commonly, these pigments owe their outstanding fastness properties to intermolecular hydrogen-bonding involving the carbonyl groups.

The quinacridones constitute one of the most important chromophoric systems developed for pigment applications since the phthalocyanines (Chapter 5). Linear *trans* quinacridones were first discovered in 1935, but their potential as pigments was not realised until the late 1950s, when they were introduced commercially by DuPont. The pigments offer outstanding fastness properties, similar to those of copper phthalocyanine, in the orange, red and violet shade areas. Structurally, quinacridones consist of a system of five fused alternate benzene and 4-pyridone rings. A number of geometrical arrangements of such a system are possible, *e.g.* **65**, **66** and **67**, but the outstanding technical properties are only given by compounds which possess the linear *trans* arrangement (**65**). Compounds **65a–c** are the products of most significant commercial importance.

At first sight, it may seem somewhat surprising that such small molecules should provide such a high degree of thermal and chemical stability and insolubility. These properties have been explained, however, by strong two-dimensional molecular association due to hydrogen bonding

65a R¹ = H; R² = H
65b R¹ = H; R² = CH₃
65c R¹ = H; R² = Cl
65d R¹ = Cl; R² = H

66

67

in the crystal structure between N–H and C=O groups, as illustrated in Figure 4.6. There is considerable evidence from related derivatives for the importance of intermolecular H-bonding in determining the properties of the quinacridones. For example, the inferior properties of the angular quinacridone **67** and the 4,11-dichloro derivative **65d** may be attributed to steric and geometric constraints, which reduce the efficiency of the hydrogen bonding. In addition, the *N,N*-dimethyl derivatives, in which no H-bonding is possible, are reported to be soluble in organic solvents. A further factor which will play a part in the technical performance of quinacridone pigments is the strong dipolar nature of the pyridone rings that arises from a major contribution from resonance forms such as **68**, leading to strong intermolecular dipolar association throughout the crystal structure.

The origin of the colour of quinacridone pigments provides a good example of the influence of crystal lattice effects, which have not yet been explained fully in fundamental terms. It is interesting that in solution the quinacridones exhibit only weak yellow to orange colours. The intense red to violet colours of the pigments in the solid state are therefore presumably due to interactions between molecules in the crystal lattice. The quinacridones show polymorphism and this has a profound effect on the colour of the pigments. The parent compound **65a** exists in three distinct polymorphic modifications each one with its own characteristic X-ray powder diffraction pattern. Two of these forms, the α- and β-modifications, are red, while the γ-form is violet. The α-form is the least stable to polymorphic change and has not been commercialised.

There is little doubt that one of the most significant developments in

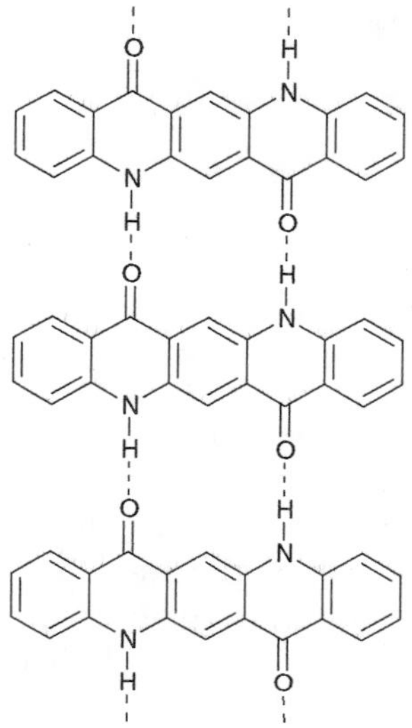

Figure 4.6 *Intermolecular association by hydrogen bonding in the crystal lattice of linear trans quinacridone,* **65a**

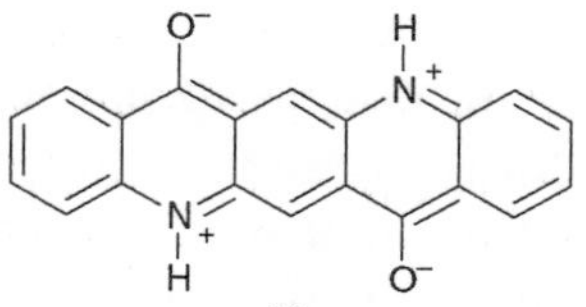

68

organic pigments in recent times is the discovery and commercialisation of pigments based on the 1,4-diketopyrrolo[3,4-*c*]pyrrole (DPP) system. The essential structural feature of this group of pigments, of which C. I. Pigment Red 254 (**69**) is a representative commercial example, are the two fused five-membered ketopyrrole rings. DPP pigments, by appropriate substituent variation, are capable of providing orange through red to bluish-violet shades. In particular, the DPP pigments which have gained most prominence provide brilliant saturated red shades of outstanding durability for automotive paint applications. In addition, their excellent thermal stability means that they are of considerable interest for the pigmentation of plastics. The pigments are, like the quinacridones, remarkable in producing such an excellent range of fastness properties from such small molecules. X-ray structural analysis has demonstrated that this is due to the strong intermolecular forces, due to hydrogen bonding and dipolar interactions, which exist throughout the crystal structure, similar to those involved with the quinacridones.

69

A number of high-grade perylene pigments, mostly reds but also including blacks, are of importance. The pigments may be represented by the general structure **70**, in which the imide nitrogen substituents, R, may be alkyl or aryl groups. An interesting observation in the perylene series is that small structural changes in the side-chain can lead to quite profound colour differences. The *N,N'*-dimethyl compound, for example, is red while the corresponding diethyl derivative is black. X-ray diffraction studies have now been applied to an extensive range of perylenes in an attempt to characterise the effect of differences in the crystal lattice structure on the light absorption properties of the pigments, a phenomenon known as crystallochromy. As an example, the *N,N'*-dimethyl compound consists of a parallel arrangement of molecules in stacks, whereas in the *N,N'*-diethyl compound the molecules are in stacks twisted with respect to one another with considerably more overlapping of the perylene ring systems in neighbouring molecules. Some perylenes with bulky substituents on the imide nitrogens are useful highly efficient and stable fluorescent dyes. They are suitable for use in demanding applications such as in solar energy collectors. The strong fluorescence of these dyes has been attributed to the structural rigidity of the molecules. A group of carbonyl pigments, which are structurally related to the perylenes are the perinones. Two isomeric perinone pigments are manufactured, C. I. Pigment Orange 43, **71**, the *trans* isomer, and C. I. Pigment Red 194, **72**, the *cis* isomer, the former being an especially important high performance product particularly for plastics applications.

THE QUINONE–HYDROQUINONE REDOX SYSTEM

The quinone–hydroquinone system represents a classic example of a fast, reversible redox system. This type of reversible redox reaction is characteristic of many inorganic systems, such as the interchange between oxidation states in transition metal ions, but it is relatively uncommon in organic chemistry. The reduction of benzoquinone to hydroquinone formally involves the transfer of two hydrogen atoms. In practice, the reversible reaction involves the stepwise transfer of two electrons from the reducing agent as shown in Scheme 4.1. Commonly, these redox reactions are carried out in alkaline media, in which case the unprotonated semi-quinone is first formed and the dianion is the final reduction product as illustrated in the scheme. In media where protons are available a series of acid–base equilibria are also involved. A wide range of dicarbonyl compounds undergo this type of reaction, including the anthraquinones, indigoid derivatives and perinones. The reduction–oxidation of carbonyl dyes is used practically in the chemistry of vat dyeing of cellulosic fibres (Chapter 7). In this process, the insoluble colorant is treated with a reducing agent in an alkaline medium to give the reduced or 'leuco' form. After application of the 'leuco' form of the dye to the fibre, the process is reversed and the coloured dye is generated by oxidation. The reducing agents commonly used include sodium dithionite ($Na_2S_2O_4$) and sodium hydrogen sulfite, while appropriate oxidising agents include sodium dichromate, iron(III), atmospheric oxygen and hydrogen peroxide.

SYNTHESIS OF CARBONYL COLORANTS

The synthesis of carbonyl colorants uses a wide diversity of chemical methods, in which each individual product essentially has its own characteristic route. This is in complete contrast to the synthesis of azo dyes and pigments (Chapter 3) where a common reaction sequence is universally used. The subject is too vast to attempt to be comprehensive in a text of this type. The following section, therefore, presents an overview of some of the fundamental synthetic strategies which may be used to prepare some of the more important types of carbonyl colorants.

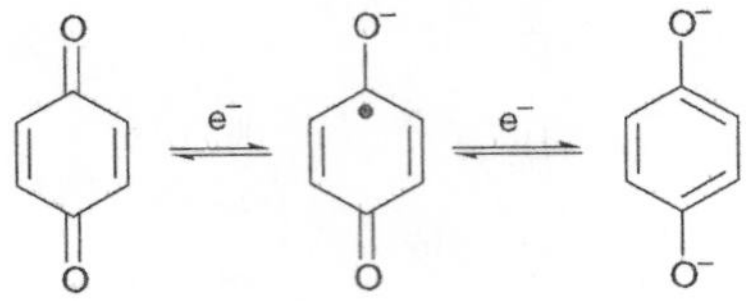

Scheme 4.1   *The benzoquinone–hydroquinone redox system*

Synthesis of Anthraquinones

The synthesis of anthraquinone colorants may effectively be envisaged as involving two stages. The first stage involves the construction of the anthraquinone ring system and in the second phase the anthraquinone nucleus is elaborated to produce the desired structure. Frequently the latter involves substitution reactions, but group interconversion and further cyclisation reactions may also be employed. Although the chemistry of the synthesis of most anthraquinone dyes and pigments is long established, some of the mechanistic detail of the individual reactions remains unexplained.

There are two principal ways of constructing the anthraquinone system which are successful industrially. The first of these is the oxidation of anthracenes and the second involves a Friedel–Crafts acylation route. Anthracene (**73**), a readily available raw material since it is a major constituent of coal tar, may be oxidised to give anthraquinone (**52**) in high yield as illustrated in Scheme 4.2. The most important oxidising agents used in this process are sodium dichromate and sulfuric acid (chromic acid) or nitric acid. This route is of considerable importance for the synthesis of the parent anthraquinone (**52**), but it is of much less importance for the direct synthesis of substituted anthraquinones. The main reasons for this are that there are few substituted anthracenes readily available as starting materials and also because many substituents would be susceptible to the strongly oxidising conditions used.

An alternative route to anthraquinone, which involves Friedel–Crafts acylation, is illustrated in Scheme 4.3. This route uses benzene and phthalic anhydride as starting materials. In the presence of aluminium(III) chloride, a Lewis acid catalyst, these compounds react to form 2-benzoyl-benzene-1-carboxylic acid, **74**. The intermediate **74** is then heated with concentrated sulfuric acid under which conditions cyclisation to anthraquinone **52** takes place. Both stages of this reaction sequence involve Friedel–Crafts acylation reactions. In the first stage the reaction is intermolecular, while the second step in which cyclisation takes place, involves an intramolecular reaction. In contrast to the oxidation route, the Friedel–Crafts route offers considerable versatility. A range of substituted

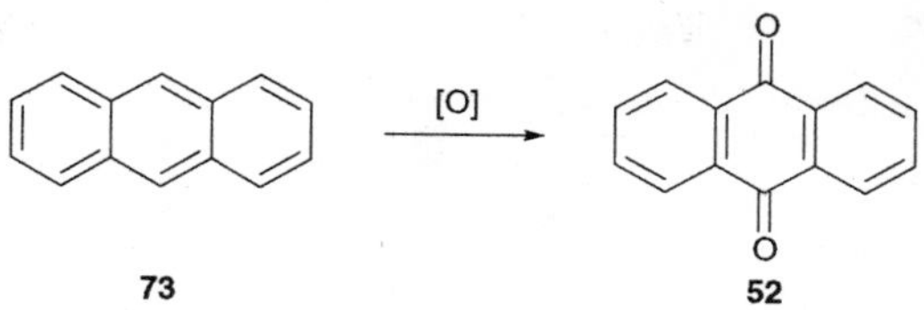

Scheme 4.2 *Oxidation of anthracene*

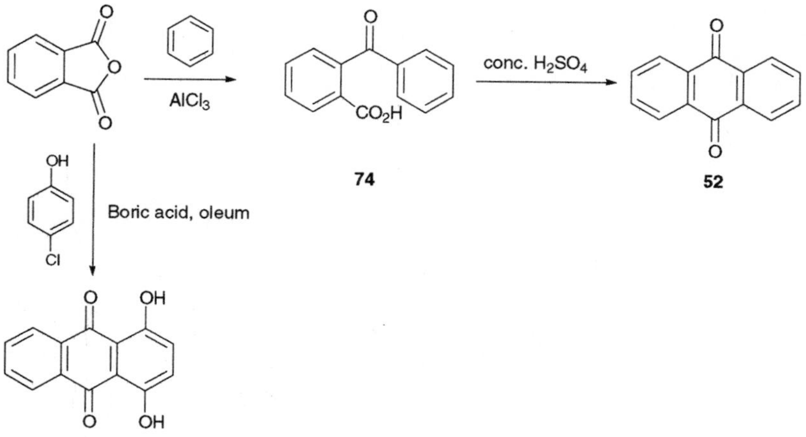

Scheme 4.3 *Friedel–Crafts route to anthraquinones*

benzene derivatives and phthalic anhydrides may be brought together as starting materials, leading to a wide range of substituted anthraquinones. For example, the use of toluene rather than benzene leads directly to 2-methylanthraquinone. A further industrially important example is also illustrated in Scheme 4.3. Starting from 4-chlorophenol and phthalic anhydride, two inexpensive and readily available starting materials, 1,4-dihydroxyanthraquinone, quinizarin, (**75**), an important intermediate in the synthesis of a number of anthraquinone dyes, may be synthesised efficiently in a 'one-pot' reaction. Boric acid in oleum is used as the catalyst and solvent system in this case. During the course of the reaction in which the anthraquinone nucleus is formed, a hydroxy group replaces the chlorine atom. The detailed mechanism by which this takes place has not been fully established.

The outer rings of the anthraquinone molecule (**52**) are aromatic in nature and as such are capable of undergoing substitution reactions. The reactivity of the rings towards substitution is determined by the fact that

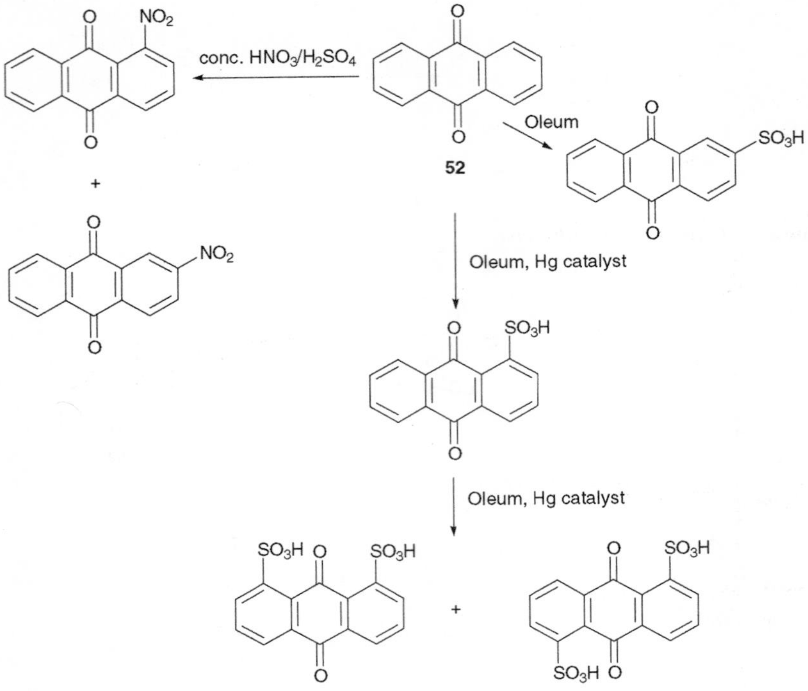

Scheme 4.4 *Some useful electrophilic substitution reactions of anthraquinone (52)*

they are attached to two electron-withdrawing carbonyl groups. The presence of these groups deactivate the aromatic rings towards electrophilic substitution. Nevertheless, using reasonably vigorous conditions, anthraquinone may be induced to undergo electrophilic substitution reactions, notably nitration and sulfonation. Scheme 4.4 illustrates a series of electrophilic substitution reactions of anthraquinone. Nitration of anthraquinone with mixed concentrated nitric and sulfuric acids leads to a mixture of the 1- and 2-nitroanthraquinones which is difficult to separate and so this process is of limited use. However, nitration of some substituted anthraquinones, particularly some hydroxy derivatives, can give more useful products. Sulfonation of anthraquinone requires the use of oleum at elevated temperatures, and under these conditions the reaction leads mainly to the 2-sulfonic acid. However, in the presence of a mercury(II) salt as catalyst, the 1-isomer, a much more useful dye intermediate, becomes the principal product. A probable explanation for these observations is that, in the absence of a catalyst, the 2–position is preferred sterically. In the presence of the catalyst, it has been proposed that mercuration takes place first preferentially at the 1–position, and the mercury is displaced subsequently by the electrophile responsible for

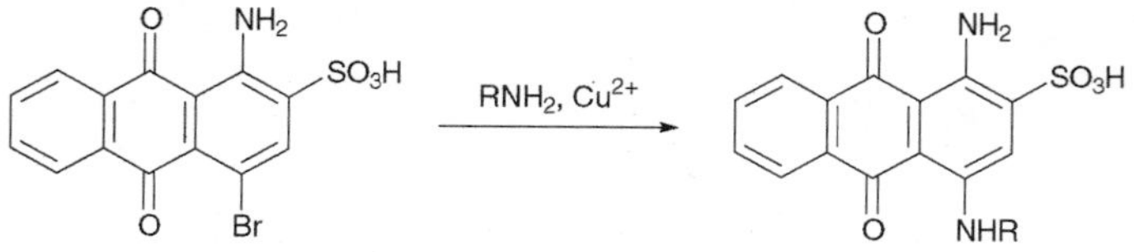

Scheme 4.5 *Reaction of bromamine acid with aromatic amines*

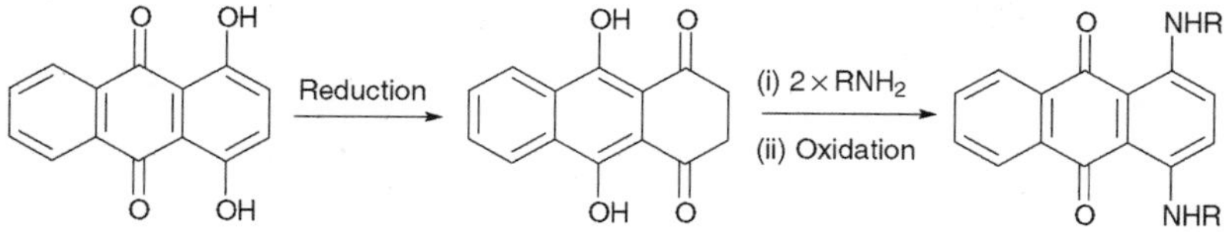

Scheme 4.6 *Formation of 1,4-diaminoanthraquinones from quinizarin*

sulfonation. Disulfonation of anthraquinone in the presence of mercury salts leads to a mixture of the 1,5- and 1,8-disulfonic acids, which are easily separated and these are also useful dye intermediates.

Nucleophilic substitution reactions, to which the aromatic rings are activated by the presence of the carbonyl groups, are commonly used in the elaboration of the anthraquinone nucleus, particularly for the introduction of hydroxy and amino groups. Commonly these substitution reactions are catalysed by either boric acid or by transition metal ions. As an example, amino and hydroxy groups may be introduced into the anthraquinone system by nucleophilic displacement of sulfonic acid groups. Another example of an industrially useful nucleophilic substitution is the reaction of 1-amino-4-bromoanthraquinone-2-sulfonic acid (bromamine acid) (**76**) with aromatic amines, as shown in Scheme 4.5, to give a series of useful water-soluble blue dyes. The displacement of bromine in these reactions is catalysed markedly by the presence of copper(II) ions.

An important route to 1,4-diaminoanthraquinones, represented by structure **78**, is illustrated in Scheme 4.6. Quinizarin (**75**) is first reduced to leucoquinizarin, which has been shown to exist as the diketo structure **77**. Condensation of compound **77** with two moles of an amine, followed by oxidation leads to the diaminoanthraquinone **78**. Boric acid is a useful catalyst for this reaction, particularly when less basic amines are used.

The syntheses of three polycyclic anthraquinones, indanthrone (**53**), pyranthrone (**55a**) and flavanthrone (**55b**), are illustrated in Scheme 4.7. In spite of the structural complexity of the products, the syntheses of these types of compound are often quite straightforward, involving, for

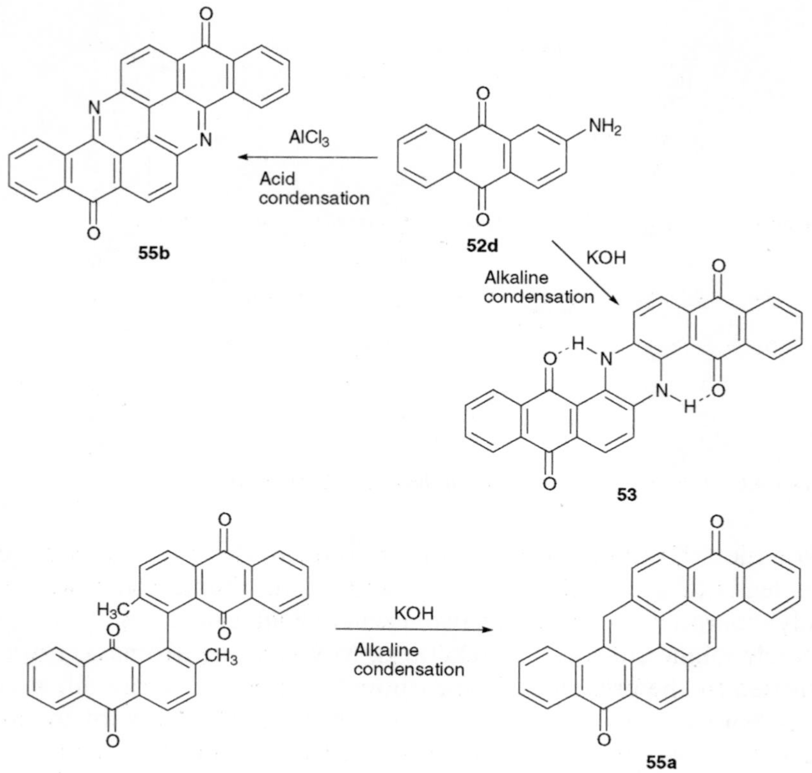

Scheme 4.7 *Syntheses of the polycyclic anthraquinones indanthrone (53), pyranthrone (55a) and flavanthrone (55b)*

example, condensation or oxidative cyclisation reactions. For example, the blue vat dye indanthrone (53) is prepared by fusion of 2-aminoanthraquinone (52d) with either sodium or potassium hydroxide at around 220 °C. Curiously, the same dye can be prepared from 1-aminoanthraquinone (52c) in a similar way, although the 2-isomer is the usual industrial starting material. Pyranthrone (55a) is also prepared by an alkaline fusion process, starting in this case from 2,2′-dimethyl-1,1′-dianthraquinonyl. The methyl groups in this molecule are sufficiently acidic as a result of activation by the carbonyl groups to be ionised and subsequently give rise to cyclisation by a reaction analogous to an aldol condensation. Flavanthrone (55b) may also be prepared by a condensation reaction starting from 2-aminoanthraquinone (52d). The yellow vat dye is obtained from fusion of 52d with alkali at temperatures of around 300 °C or by acid-catalysed condensation, for example using aluminium(III) chloride. However, its industrial manufacture usually involves a

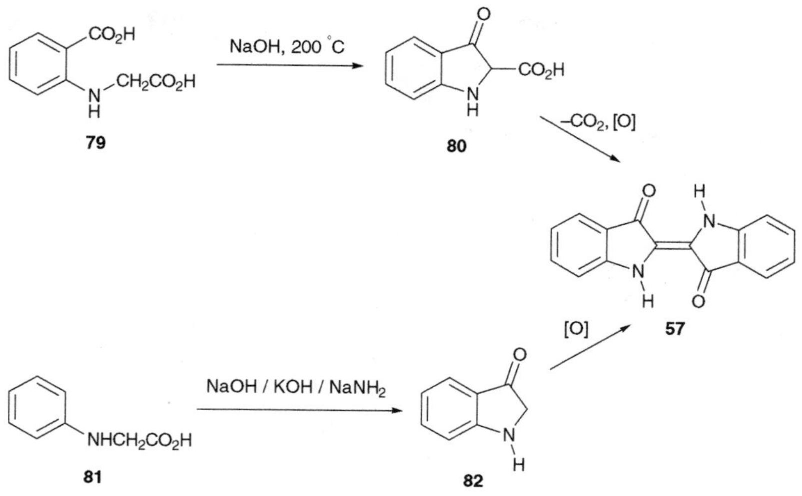

Scheme 4.8 *Synthetic routes to indigo (57)*

similar condensation process, either acid or base catalysed, starting from 1-chloro-2-acetylaminoanthraquinone.

Synthesis of Indigoid Colorants

Indigo (**57**) was for many centuries obtained from natural sources. The chemical structure of indigo was first proposed by von Baeyer in 1869, and eleven years later he reported the first successful synthesis, a multi-stage route starting from *o*-nitrocinnamic acid. The first successful commercial synthesis of indigo, attributed to Heumann in 1897, is shown in Scheme 4.8. In this classical synthesis, phenylglycine-*o*-carboxylic acid (**79**) is converted by fusion with sodium hydroxide at around 200 °C, in the absence of air, into indoxyl-2-carboxylic acid (**80**). This material readily decarboxylates and oxidises in air to indigo. A much more efficient synthesis, which forms the basis of the manufacturing method in use today, is due originally to Pfleger (1901). In this route, also illustrated in Scheme 4.8, the more readily available starting material, phenylglycine (**81**) is treated in an alkaline melt of sodium and potassium hydroxides containing sodamide. This process leads directly to indoxyl (**82**), which undergoes spontaneous oxidative dimerisation in air to indigo. Thioindigo (**60b**) is best prepared from *o*-carboxybenzenethioglycollic acid (**83**) by a route analogous to Heumann's indigo synthesis. The final oxidation step in this case uses sulfur rather than oxygen as the oxidising agent.

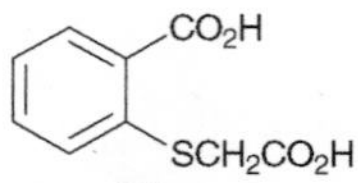

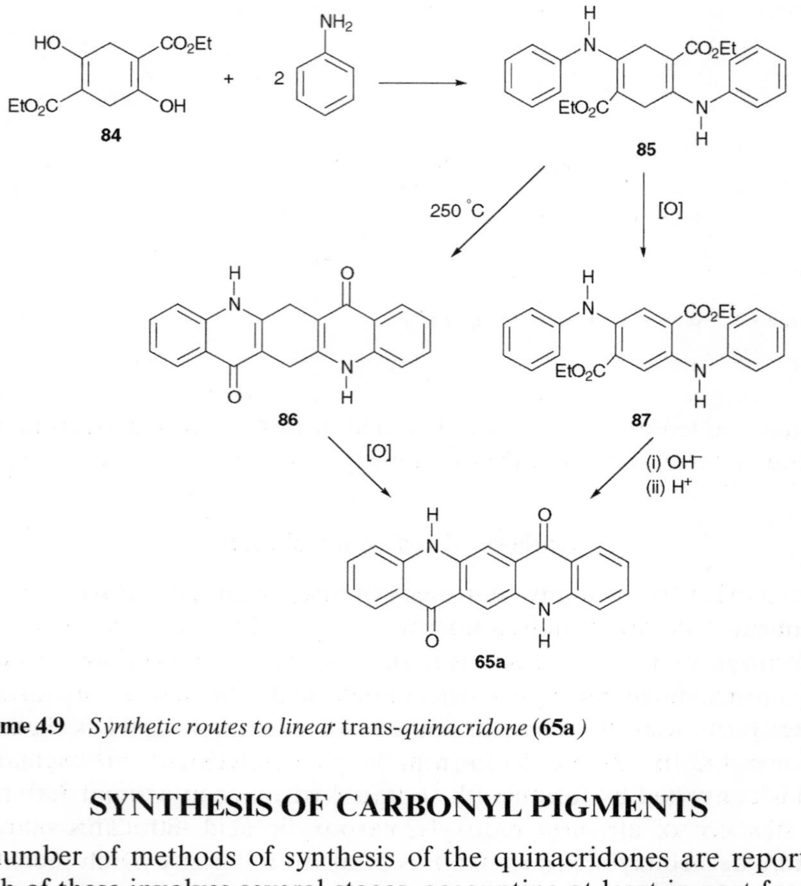

Scheme 4.9 *Synthetic routes to linear* trans-*quinacridone* (**65a**)

SYNTHESIS OF CARBONYL PIGMENTS

A number of methods of synthesis of the quinacridones are reported. Each of these involves several stages, accounting at least in part for the somewhat higher cost of these pigments. The two most important routes to the parent compound are outlined in Scheme 4.9. In both cases, the starting material, diethyl succinylsuccinate (**84**, 1 mol), which may be prepared by a base-catalysed self-condensation of a succinic acid diester, is condensed with aniline (2 mol) to form the 2,5-diphenylamino-3,6-dihydroterephthalic acid diester (**85**). Diester **85** undergoes ring closure at an elevated temperature in a high boiling solvent to give the dihydroquinacridone (**86**), which is relatively easily oxidised (*e.g.* with

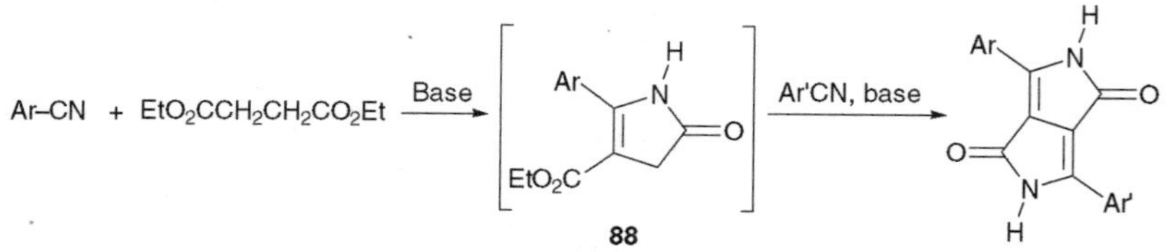

Scheme 4.10 *Synthesis of DPP pigments*

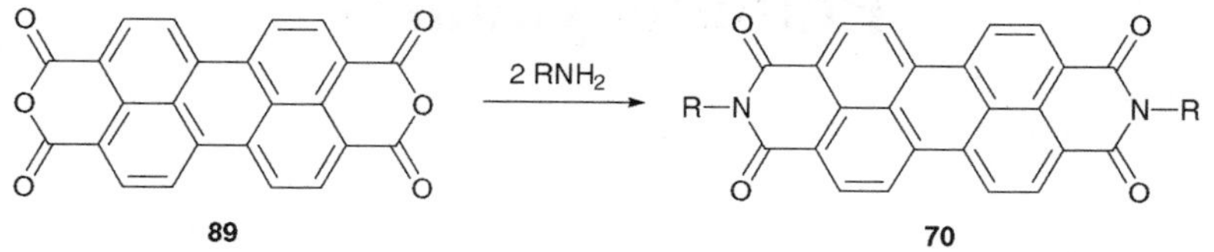

Scheme 4.11 *Synthesis of perylenes*

sodium 3-nitrobenzene-1-sulfonate) to the quinacridone **65a**. Alternatively, compound **85** may be oxidised to the 2,5-diarylaminoterephthalate diester **87**. Base hydrolysis of the diester, followed by ring closure by treatment, for example, with polyphosphoric acid gives the quinacridone.

The formation of a DPP molecule was first reported in 1974 as a minor product in low yield from the reaction of benzonitrile with ethyl bromoacetate and zinc. A fascinating study by research chemists at Ciba Geigy into the mechanistic pathways involved in the formation of the molecules led to the development of an efficient 'one-pot' synthetic procedure to yield DPP pigments from readily available starting materials, as illustrated in Scheme 4.10. The reaction involves the treatment of diethyl succinate (1 mol) with an aromatic cyanide (2 mol) in the presence of a strong base. The reaction proceeds through the intermediate **88**, which may be isolated and used to synthesise unsymmetrical derivatives.

Perylenes (**70**) are diimides of perylene-3,4,9,10-tetracarboxylic acid, and may be prepared by reaction of the bis-anhydride of this acid, **89** (1 mol) with the appropriate amine (2 mol) in a high-boiling solvent as illustrated in Scheme 4.11. The synthesis of perinones **71** and **72** involves condensation of naphthalene-1,4,5,8-tetracarboxylic acid with benzene-1,2-diamine in refluxing acetic acid. This affords a mixture of the two isomers, which may be separated by a variety of methods, generally involving their differential solubility in acids and alkalis.

Chapter 5

Phthalocyanines

The phthalocyanines represent without doubt the most important chromophoric system developed during the 20th century. Historically, the most important event was probably their accidental discovery around 1928 by a dye manufacturing company in Scotland. However, there is little doubt that researchers before this had isolated phthalocyanines, but the significance of their observations was not fully recognised. In 1907, Braun and Tcherniac were engaged in a study of the chemistry of *o*-cyanobenzamide (**90**) and discovered that when this compound was heated a trace amount of a blue substance was obtained. This compound undoubtedly was metal-free phthalocyanine (**91**). In 1927, de Diesbach and co-workers reported that when 1,2-dibromobenzene was treated with copper(I) cyanide in boiling quinoline for eight hours, a blue product was obtained in reasonable yield. This was almost certainly the first preparation of copper phthalocyanine (CuPc, **92**). They obtained the molecular formula of the compound from elemental analysis and noted its remarkable stability to alkali, concentrated acids and heat but were unable to propose a structure. In 1928, in the manufacture of phthalimide by Scottish Dyes (later to become part of ICI) from the reaction of phthalic anhydride with ammonia in a glass-lined reactor, the formation of a blue impurity was observed in certain production batches. This contaminant was isolated as a dark blue, insoluble crystalline substance. Ultimately, the compound proved to be iron phthalocyanine (FePc), the source of the iron being the wall of the reactor which became exposed due to a flaw in the glass lining. An independent synthesis involving passing ammonia gas through molten phthalic anhydride in the presence of iron filings confirmed the findings. Following this discovery, the colour manufacturing industry was quick to recognise the unique properties of the compounds and to exploit their commercial potential. The phthalocyanines have emerged as one of the most extensively studied classes of compounds, because of their intense, bright colours, their high stability and their unique molecular structure.

THE STRUCTURE AND PROPERTIES OF PHTHALOCYANINES

The elucidation of the structure of the phthalocyanines followed some pioneering research into the chemistry of the system by Linstead of Imperial College, University of London. The structure that we now recognise was first proposed from the results of analysis of a number of metal phthalocyanines, which provided the molecular formulae, and from an investigation of the products from degradation studies. Finally, Robertson confirmed the structure as a result of one of the classical applications of single crystal X-ray crystallography.

The phthalocyanine system, which may be considered as the tetraaza derivative of tetrabenzoporphin, is planar, consisting of four isoindole units connected by four nitrogen atoms that form together an internal 16-membered ring of alternate carbon and nitrogen atoms. Most phthalocyanines contain a central complexed metal atom, derivatives having been prepared from the majority of the metals in the periodic table. The central metal atom is in a square planar environment. The phthalocyanines are structurally related to the natural pigments chlorophyll (**93**) and haemin (**94**), which are porphyrin derivatives. However, unlike these natural colorants, which have poor stability, the phthalocyanines exhibit exceptional stability and they are in fact probably the most stable of all synthetic organic colorants. Copper phthalocyanine, used here as an example, is usually illustrated as structure **92**, which contains three benzenoid and one *o*-quinonoid outer rings. However, it has been established that the molecule is centrosymmetric and this means that structure **92** should be regarded as only one of a large number of resonance forms contributing to the overall molecular structure. The extensive resonance stabilisation of the phthalocyanines may well account for their high stability. The phthalocyanines are aromatic

93

R = CH₃: chlorophyll a
R = CHO: chlorophyll b

94

molecules, a feature that has been attributed to the 18 π-electrons in the perimeter of the molecules.

The metal phthalocyanines in general show brilliant, intense colours. The UV/visible absorption spectrum of metal-free phthalocyanine (**91**) in 1-chloronaphthalene shows two absorption bands of similar intensity at 699 and 664 nm. The corresponding spectra of metal phthalocyanines, however, show a single narrow major absorption band, a feature which has been explained by their higher symmetry compared with the metal-free compound and it is the nature of this absorption which gives rise to the brilliance and intensity of their colour. The colours of traditional phthalocyanine dyes and pigments are restricted to blues and greens, although recent years have seen the development of a number of derivatives whose absorption is extended into the near-infrared region of the spectrum. Phthalocyanines are of the polyene rather than the donor–acceptor chromogenic type. As a consequence, the valence-bond (resonance) approach may not be applied so readily to provide an explanation of their colour (Chapter 2). In contrast, the phthalocyanines and structurally related systems have been investigated extensively by molecular orbital methods, including the PPP approach, and these methods generally provide a successful account of the light absorption properties of the system.

The position of the absorption band of metal phthalocyanines is dependent on the nature of the central metal ion, the substituent pattern on the outer rings and the degree of ring annelation. Among the most extensively investigated phthalocyanines are the complexes of the first transition series metals, iron, cobalt, nickel, copper and zinc. Within this series, the colour is affected little by the nature of the central metal ion; the λ_{max} values are to be found in the range 670–685 nm. The most hypsochromic of the series of unsubstituted metal phthalocyanines is PtPc (λ_{max} 652 nm) while the most bathochromic is PbPc (λ_{max} 714 nm). Neither of these is of particular interest commercially due either to economic or toxicity considerations, but the bathochromic effect of the vanadyl derivative, VOPc, (λ_{max} 701 nm) is of interest from the point of view of extending the absorption range of phthalocyanines. Substituents on the outer aromatic rings almost invariably shift the absorption band to longer wavelengths. Copper hexadecachlorophthalocyanine, for example, absorbs at 720 nm in 1-chloronaphthalene, giving rise to its green colour. There is some considerable interest in phthalocyanines in which the absorption band is extended into the near-infrared region for applications such as optical data storage and security printing (Chapter 10). This may be achieved in a number of ways. For example, the arylthio group causes a much more pronounced bathochromic shift than the halogens and a number of copper polyarylthiophthalocyanine derivatives have been patented for applications which make use of their intense absorption in the near-infrared region of the spectrum. Alternatively, extending the outer ring system by annelation shifts the absorption band bathochromically. Copper naphthalocyanine (**95**), for example, absorbs at 784 nm, while the corresponding vanadyl derivative gives a λ_{max} value of 817 nm.

95

While textile dyes based on phthalocyanines are of rather limited importance, the phthalocyanines provide by far the most important blue and green organic pigments. In particular, copper phthalocyanine (**92**), C. I. Pigment Blue 15, is by far the most important blue pigment, finding almost universal use as a colorant in a wide range of paint, printing ink and plastics applications. In fact, there is a convincing argument that it is the most important of all organic pigments. It owes this dominant position to its intense brilliant blue colour and excellent technical performance. The pigment exhibits exceptional stability to light, heat, solvents, alkalis, acids and other chemicals. Among the features that demonstrate this high stability are the ability of the material to sublime unchanged at temperatures above 500 °C, and the observation that it dissolves without decomposition in concentrated sulfuric acid, from which solutions it may be recovered. In addition, copper phthalocyanine is a relatively low cost product since, in spite of its structural complexity, its manufacture (see next section) is straightforward, giving high yields from inexpensive, commodity starting materials.

Copper phthalocyanine exhibits polymorphism, the most important crystal phases being the α- and β-forms, and at least three other forms have been reported. Both the α- and β-forms are of commercial import-ance. The two forms exhibit different hues, the α-form being reddish-blue while the β-form is greenish-blue. The β-form is the more stable form, particularly towards organic solvents. The α-form has a tendency to convert in the presence of certain solvents into the β-form with a corre-sponding change in shade, unless it is stabilised, for example by the incorporation of a single chlorine substituent. β-CuPc is of particular importance as the cyan pigment used most commonly in printing inks while the α-form is more important in surface coatings and plastics applications. A structural comparison between the α- and β-phases is shown in Figure 5.1. It has been suggested that in the 'herring-bone' arrangement of CuPc molecules in the crystal structure of the β-form the copper atom at the centre of each molecule is coordinated to nitrogen atoms in adjacent molecules, forming a distorted octahedron, a coordina-tion geometry which is particularly favoured in complexes of copper. No such octahedral coordination is possible in the parallel arrangement of molecules in the crystal structure of α-CuPc, a factor which may contrib-ute to the lower stability of this polymorphic form.

A number of other phthalocyanines are used commercially as pig-ments. The most important green organic pigments are the halogenated copper phthalocyanines C. I. Pigment Green 7, in which the 16 ring

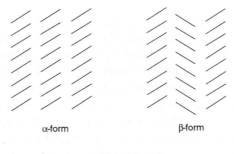

α-form   β-form

——— Planar CuPc molecules

Figure 5.1   *The polymorphism of copper phthalocyanine*

hydrogen atoms of the CuPc molecule are replaced virtually completely by chlorine, and C. I. Pigment Green 36, a designation which incorporates a range of bromo- and bromochloro-copper phthalocyanines. The hue of these pigments becomes progressively yellower with increasing bromine substitution. The phthalocyanine greens exhibit the same outstanding colouristic and technical performance as the blue pigments from which they are derived and find equally widespread use in the coloration of paints, printing inks and plastics. Although phthalocyanine complexes have been prepared from virtually every metallic element in the periodic table, only the copper derivatives are of significant commercial importance as pigments, simply because the copper compounds give the best combination of colour and technical properties. However, metal-free phthalocyanine (**91**) finds some use as a greenish-blue pigment of high stability.

In view of the immense commercial importance of phthalocyanines as pigments, it is perhaps surprising that only a few are of importance as textile dyes. This is primarily due to the size of the molecules; they are too large to allow penetration into many fibres, especially the synthetic fibres polyester and polyacrylonitrile. An example of a phthalocyanine dye which may be used to dye cellulosic substrates such as cotton and paper is C. I. Direct Blue 86 (**96**), a disulfonated copper phthalocyanine. In addition, a few blue reactive dyes for cotton incorporate the copper phthalocyanine system as the chromophoric unit (Chapter 8).

96

SYNTHESIS OF PHTHALOCYANINES

The synthesis of metal phthalocyanines requires, essentially, the presence of three components: a phthalic acid derivative, such as phthalic anhydride, phthalimide, phthalonitrile or *o*-cyanobenzamide, a source of nitrogen (in cases where the phthalic acid derivative does not itself contain sufficient nitrogen) and an appropriate metal derivative. Commonly the reaction requires high temperatures and may be carried out in a high boiling solvent or as a 'dry bake' process. In this way, using appropriate starting materials and reaction conditions, virtually the entire range of

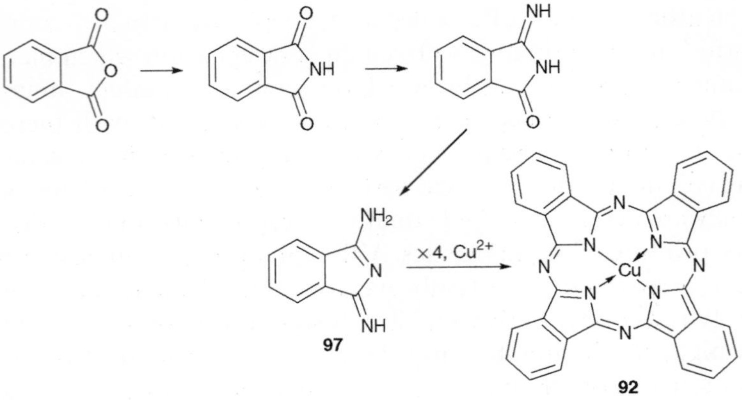

Scheme 5.1 *The phthalic anhydride route to CuPc*

metal phthalocyanines may be prepared. Substituted phthalocyanines are prepared either by using an appropriately substituted phthalic acid derivative as a starting material, or by substitution reactions carried out on the unsubstituted derivatives. Metal-free phthalocyanines are conveniently prepared by subjecting certain labile metal derivatives, such as those of sodium or lithium, to acidic conditions. The method of synthesis is discussed further in this section for the case of copper phthalocyanine, because of the particular importance of this product. Although the structure of copper phthalocyanine is rather complex, its synthesis is remarkably straightforward. It may be prepared in virtually quantitative yield from readily available, low cost starting materials. Two chemically related methods, the phthalic anhydride and phthalonitrile routes, are commonly used for its manufacture. Both involve simultaneous synthesis of the ligand and metal complex formation in a template procedure.

(a) The phthalic anhydride route. In the most commonly encountered version of this method, phthalic anhydride is heated with urea, copper(I) chloride and a catalytic amount of ammonium molybdate in a high-boiling solvent. An outline of the process is given in Scheme 5.1. Urea acts as the source of nitrogen in the process, the carbonyl group of the urea molecule being displaced as carbon dioxide. Mechanistic schemes have been proposed to explain the course of this synthesis but much of the detail remains to be established unequivocally. In essence, phthalic anhydride reacts with urea or products of its decomposition or polymerisation, resulting in progressive replacement of the oxygen atoms by nitrogen and, ultimately, the formation of the key intermediate 1-amino-3-iminoisoindoline (**97**). The presence of ammonium molybdate is essential

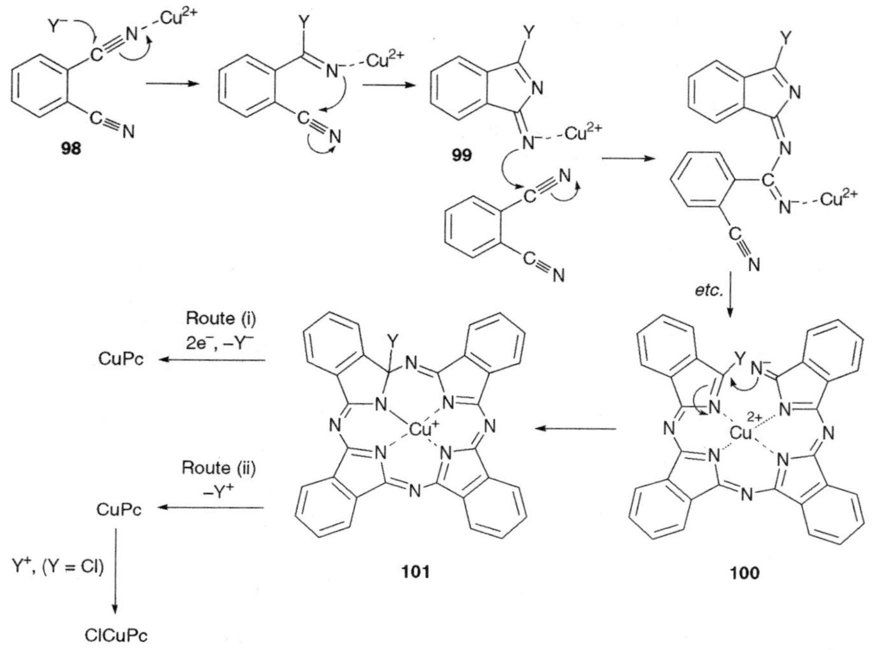

Scheme 5.2 *A mechanism for the phthalonitrile route to CuPc*

to catalyse this part of the sequence. Subsequently, this intermediate undergoes a tetramerisation with cyclisation aided by the presence of the copper ion to form copper phthalocyanine.

(b) The phthalonitrile route. In this process, phthalonitrile (**98**) is heated to around 200 °C with copper metal or a copper salt, with or without a solvent. A mechanism for the phthalonitrile route to copper phthalocyanine has been proposed as illustrated in Scheme 5.2. It is suggested that reaction is initiated by attack by a nucleophile (Y^-), most likely the counter-anion associated with the Cu^{2+} ion, at one of the cyano groups of the phthalonitrile activated by its coordination with the Cu^{2+} ion. Cyclisation to isoindoline derivative **99** then takes place. Attack by intermediate **99** on a further molecule of phthalonitrile then takes place and, following a series of similar reactions, including a cyclisation step, facilitated by the coordinating role of the Cu^{2+}, intermediate **101** is formed. When copper metal is the reactant, it is proposed that two electrons are transferred from the metal, allowing elimination of Y^- to form copper phthalocyanine [route (i)]. Consequently, the Cu(0) is oxidised to Cu(II) as required to participate further in the reaction. When a copper(II) salt is used, it is suggested that Y^+ (the chloronium ion in the

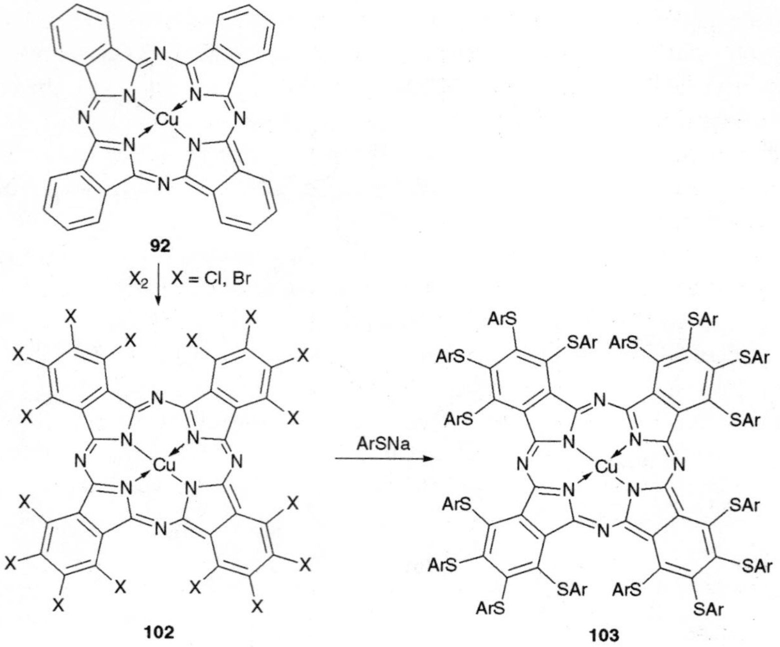

Scheme 5.3 *Reactions leading to some substituted CuPc derivatives*

case of CuCl$_2$) is eliminated to form CuPc [route (ii)]. The product in this case, rather than copper phthalocyanine itself, is a monochloro derivative formed by electrophilic attack of Cl$^+$ on the copper phthalocyanine initially formed. Copper monochlorophthalocyanine is important as it exists exclusively in the α-crystal form which, unlike unsubstituted CuPc, is stable to solvents. It has been suggested that the single chlorine atom sterically prevents conversion into the β-form.

Both the phthalic anhydride and phthalonitrile routes produce a crude blue product, which is of far too large a particle size to be of use as a pigment. The original method developed for particle size reduction used acid pasting, which involves dissolving the crude product in concentrated sulfuric acid, followed by reprecipitation with water. This method gives the α-form of the pigment in a fine particle size form. Mechanical grinding of the crude blue product in the presence of inorganic salts, such as sodium chloride or calcium chloride, produces a mixture of the α- and β-CuPc which may be converted into pure pigmentary β-CuPc by careful treatment of this mixture with certain organic solvents. Alternatively, grinding the crude material with inorganic salts in the presence of organic solvents can lead directly to β-CuPc in a fine particle size form.

Scheme 5.3 shows an outline of some important substitution reactions of copper phthalocyanine. Synthesis of the phthalocyanine green pigments involves the direct exhaustive halogenation of crude copper phthalocyanine blue with chlorine or bromine or an appropriate mixture of the two halogens, depending on the particular product required, at elevated temperatures in a suitable solvent, commonly an $AlCl_3/NaCl$ melt. These reactions are examples of electrophilic substitution, reflecting the aromatic character of the copper phthalocyanine molecule. The crude green products **102**, which are initially formed under these manufacturing conditions, are of large particle size due mainly to a high degree of aggregation. The crude form may be converted into an appropriate pigmentary form either by treatment with suitable organic solvents or by treatment with aqueous surfactant solutions. These processes effect a deaggregation of the product, producing a finer particle size, and in addition they increase the crystallinity of the products. Both of these effects provide a dramatic beneficial effect on the performance of the products as pigments. Treatment of polyhalogenated copper phthalocyanines **102** with thiophenols in the presence of alkali at high temperatures in high-boiling solvents gives the near-infrared absorbing polyarylthio CuPc derivatives **103**, as illustrated in Scheme 5.3. This process provides an example of aromatic nucleophilic substitution in the phthalocyanine system. X-ray structural analysis of these arylthio derivatives demonstrates that the sulfur atoms are located in the plane of the CuPc system while the aryl groups are twisted to accommodate the steric congestion.

Chapter 6

Miscellaneous Chemical Classes of Organic Dyes and Pigments

The chemistry of the three most important chemical classes of organic colorants, the azo, carbonyl and phthalocyanine classes, has been dealt with individually in Chapters 3–5 respectively. In this chapter, the chemistry of a further five chemical classes which are of some importance for specific applications is discussed. These classes are the polymethines, arylcarbonium ion colorants, dioxazines, sulfur dyes and nitro dyes. A section of this chapter is devoted to each of these, each individual section contains a description of the principal structural features which characterise the particular colorant type, together with an outline of the chemistry of the main synthetic routes. There are many other chemical types of dyes and pigments that do not fall into the categories previously mentioned, but which are neglected in this text either because they are commercially of little importance or because they have been less extensively investigated.

POLYENE AND POLYMETHINE DYES

Polyene and polymethine dyes are two structurally related groups of dyes which contain as their essential structural feature one or more methine (–CH=) groups. Polyene dyes contain a series of conjugated double bonds, usually terminating in aliphatic or alicyclic groups. They owe their colour therefore simply to the presence of the conjugated system. In polymethine dyes, electron-donor and electron-acceptor groups terminate either end of the polymethine chain, so that they may be considered as typical donor–acceptor dyes.

The best-known group of polyene dyes is the carotenoids, which are widely encountered natural colorants. β-Carotene (**104**) is given as an

102

104

important example. This dye is a hydrocarbon that contains eleven conjugated double bonds, which illustrates the length of chain that is necessary to shift the absorption from the UV into the visible range. Dye **104** shows absorption maxima at 450 and 478 nm. Naturally occurring carotenoids not only provide the attractive yellow, orange and red colours of many plants, but also serve important biochemical functions. For example, they may protect cells and organisms against some of the harmful effects of exposure to light as a result of their action as efficient deactivators of singlet oxygen, and they are also important in some biological energy transfer processes. Nutritionists have always encouraged us to include significant quantities of fresh fruit and vegetables in our diet. Special emphasis has been added to this advice in recent years as evidence has emerged for the potential therapeutic benefits of carotenoids and other related natural colorants, including the possibility that they may inhibit tumour development, no doubt related to their ability to quench singlet oxygen. Another carotenoid which is of some considerable functional importance is retinal, which is the chromogenic part of the molecules of rhodopsin and iodopsin, the pigments in the eye which are responsible for colour vision (Chapter 2). There are no true synthetic polyene dyes of any real commercial importance. However, the phthalocyanines (Chapter 5) may, in a sense, be considered, structurally, as aza analogues of a cyclic polyene system.

In polymethine dyes, electron donor (D) and acceptor (A) groups terminate the polymethine chain as illustrated by the general structure given in Figure 6.1. They embrace a wide variety of structural types which may be subdivided into three broad categories as either cationic ($z = + 1$), anionic ($z = - 1$) or neutral ($z = 0$) types, depending on the precise nature of A and D. Dyes of a similar structure in which one or more of the methine carbon atoms are replaced by aza nitrogen atoms are also conveniently considered as polymethines. Polymethine dyes are capable of providing a wide range of bright, intense colours but, in general, they have tended to show rather poor fastness properties compared with other chemical classes. This feature has limited their use on textiles, where they are restricted mainly to some disperse dyes for polyester and cationic dyes for acrylic fibres and basic dyes They are essentially of no industrial consequence as pigments. They are, however, used in the

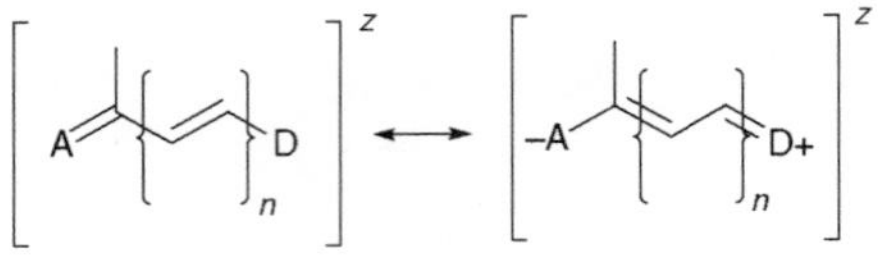

n = 0, 1, 2, *etc.*
z = −1, 0, +1

Figure 6.1 *General structure of polymethine dyes*

colour photographic industry, and they have been extensively studied because of fundamental theoretical interest.

The most important group of cationic polymethine dyes contains nitrogen atoms in both the D and A groups. These are further classified as cyanines, **105**, hemicyanines, **106**, or streptocyanines, **107**, depending on whether both, one or neither of the nitrogen atoms is contained in a heterocyclic ring, as illustrated in Figure 6.2. The history of this type of dye dates from 1856, the same year in which Perkin discovered Mauveine (Chapter 1). In that year, Williams discovered that the reaction of crude quinoline, which fortuitously contained substantial quantities of 4-methylquinoline, with isopentyl iodide in the presence of alkali gave a blue dye which he named *Cyanine*. This dye was eventually identified as having the structure **108**. A range of cyanine dyes was subsequently prepared, but these early products proved to be of little use for textiles because of their poor lightfastness. However, considerable interest in their chemistry led ultimately to the discovery that certain of the dyes had photosensitising properties and this gave rise to their application in colour photography. A new era for cationic polymethine dyes was heralded by the introduction in the 1950s of polyacrylonitrile fibres. It was found that such dyes when applied to acrylic fibres containing anionic sites were capable of giving bright, strong shades with good fastness properties (Chapter 7). Examples of dyes of this type are the cyanine C. I. Basic Red 12, **109**, the hemicyanine C. I. Basic Violet 7, **110**, the diazahemicyanine C. I. Basic Blue 41, **111** (which may also be considered as a member of the azo dye class), the azacyanine C. I. Basic Yellow 11, **112**, and the diazacyanine C. I. Basic Yellow 28, **113**. In this series, the cyanines and their aza analogues generally give yellow through to red dyes, while the corresponding hemicyanines are more bathochromic, giving reds, violets and blues.

The structures of a number of neutral and anionic polymethine dyes are illustrated in Figure 6.3. There are many types of neutral polymethines, utilising a wide range of electron donor and acceptor groups. For example, C. I. Disperse Yellow 99 (**114**) and C. I. Disperse Blue 354

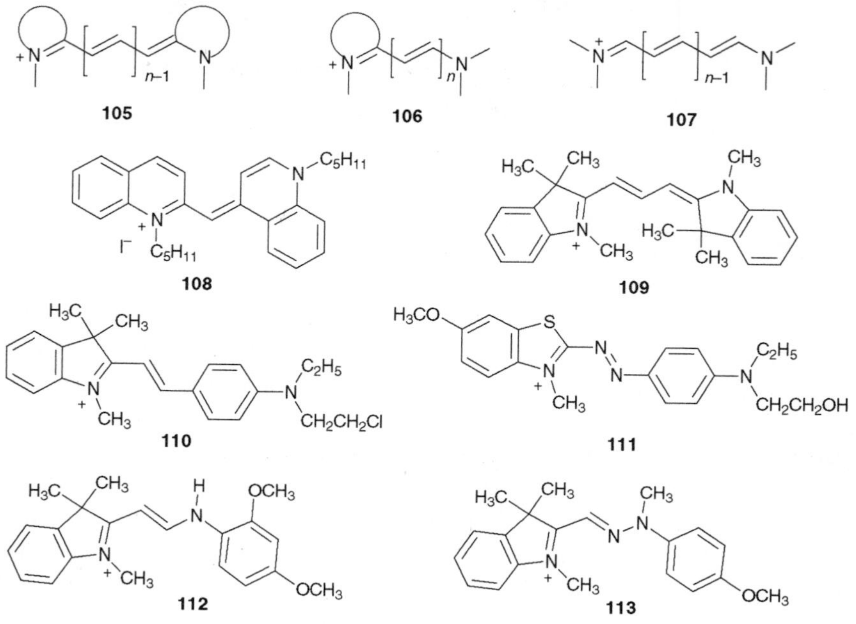

Figure 6.2   *Structures of some cationic polymethine dyes*

(**115**) are important dyes for polyester (Chapter 7). Merocyanines in which the amino group is the donor and the carbonyl group is the acceptor, as represented by the general structure **116**, are well known. Because of their generally inferior lightfastness properties they are not used on textiles, but find wider use in colour photography. Merocyanines are also formed when certain colourless photochromic compounds, notably the spirooxazines, are irradiated with light. The reversible colour change given by these compounds may be used in a range of applications, for example in ophthalmics, security printing and optical data storage (Chapter 10). Anionic polymethines, such as the oxonol, **117**, are usually rather unstable components and so they have not been so extensively investigated.

Symmetrical cyanine dyes, because of the resonance shown in Figure 6.4 (in which the two contributing structures are exactly equivalent), are completely symmetrical molecules. X-ray crystal structure determinations and NMR spectroscopic analysis have demonstrated that the dyes are essentially planar and that the carbon–carbon bond lengths in the polymethine chain are uniform. The colour of cyanine dyes depends mainly on the nature of the terminal groups and on the length of the polymethine chain. The bathochromicity of the dyes is found to increase

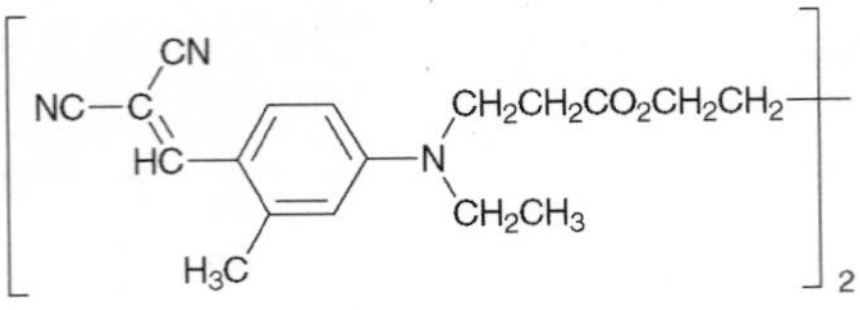

114 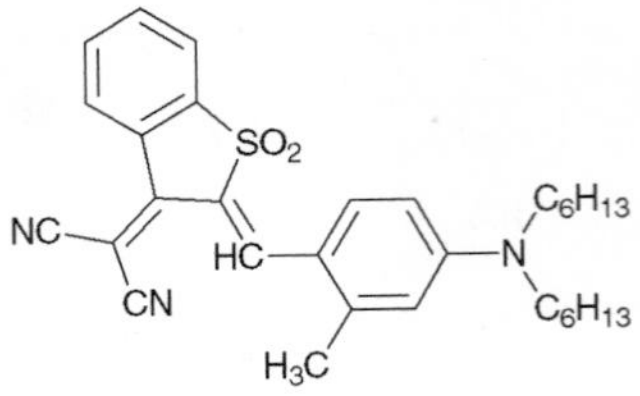   **115**

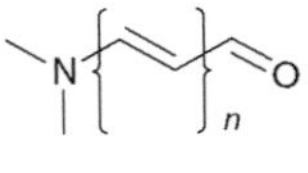

116 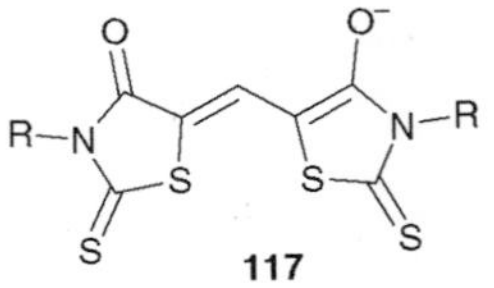   **117**

Figure 6.3   *Structures of some neutral and anionic polymethine dyes*

Figure 6.4   *Valence-bond (resonance) approach to cyanine dyes*

with the electron-releasing power of the terminal donor group. More bathochromic dyes are also obtained by incorporating the terminal group into a heterocyclic ring system and by extending the conjugation. Cyanine dyes display a highly allowed HOMO → LUMO transition, and hence they show high molar extinction coefficients, *i.e.* high colour strength. Another important feature of cyanine dyes is the narrowness of the absorption bands, which means that they are exceptionally bright colours. One factor which influences the absorption bandwidth of a dye is how closely the geometry of the first excited state of the molecule resembles that of the ground state. In the case of cyanine dyes these two states exhibit very similar geometry and hence the absorption bands are narrow. The results of molecular orbital calculations, for example using the PPP-MO approach, are in good agreement with the experimental spectral data for a wide range of polymethine dyes. In addition, the calculations confirm that there is no major redistribution of π-electron charge densities or π-bond orders on excitation.

There has been some interest in extending the absorption range of cyanine dyes to longer wavelengths into the near-infrared region of the spectrum. Consideration of the spectral data for thiazole derivatives **118–120** is of some interest in this respect. Cyanine dye **118** shows the characteristic visible absorption spectrum for a dye of this type, giving a

118

119

120

narrow band with a λ_{max} value of 651 nm in acetonitrile. As the length of the conjugated polymethine chain of cyanine dye **118** is extended further, the absorption band is shifted bathochromically by about 100 nm for each additional $-CH=CH-$ group. However, the absorption curves become increasingly broad as the chain is extended and there is a consequent reduction in the molar extinction coefficient. It has been suggested that this may be due to *trans–cis* isomerism, which becomes more facile as the length of the chain increases, and also to an increase in the participation of higher vibrational states of the first excited state as the molecular flexibility is increased. As an alternative to simple extension of the conjugation some structurally related systems have been investigated. The squarylium dye **119** absorbs at 663 nm and gives a considerably narrower bandwidth than compound **118**, while the croconium derivative **120** gives a narrow absorption band at 771 nm. Dyes of these structural types are of some interest for their potential to provide intense narrow absorption bands in the near-infrared region with minimal absorption in the visible region (Chapter 10).

The strategies used in the synthesis of polymethine dyes are illustrated for a series of indoline derivatives in Scheme 6.1. There is an even wider range of synthetic routes to polymethine dyes than is described here, but they are based for the most part on a similar set of principles. The starting material for the synthesis of this group of polymethine dyes is invariably 2-methylene-1,3,3-trimethylindolenine (**121**), known universally as Fischer's base. As illustrated in the scheme, compound **121** may be converted by formylation using phosphoryl chloride and dimethylformamide into compound **122**, referred to as Fischer's aldehyde, which is also a useful starting material for this series of polymethine dyes. When compound **121** (2 mol) is heated with triethylorthoformate (1 mol) in the presence of a base such as pyridine, the symmetrical cyanine dye, C. I. Basic Red 12 **109** is formed. The synthesis of some hemicyanines may be achieved by

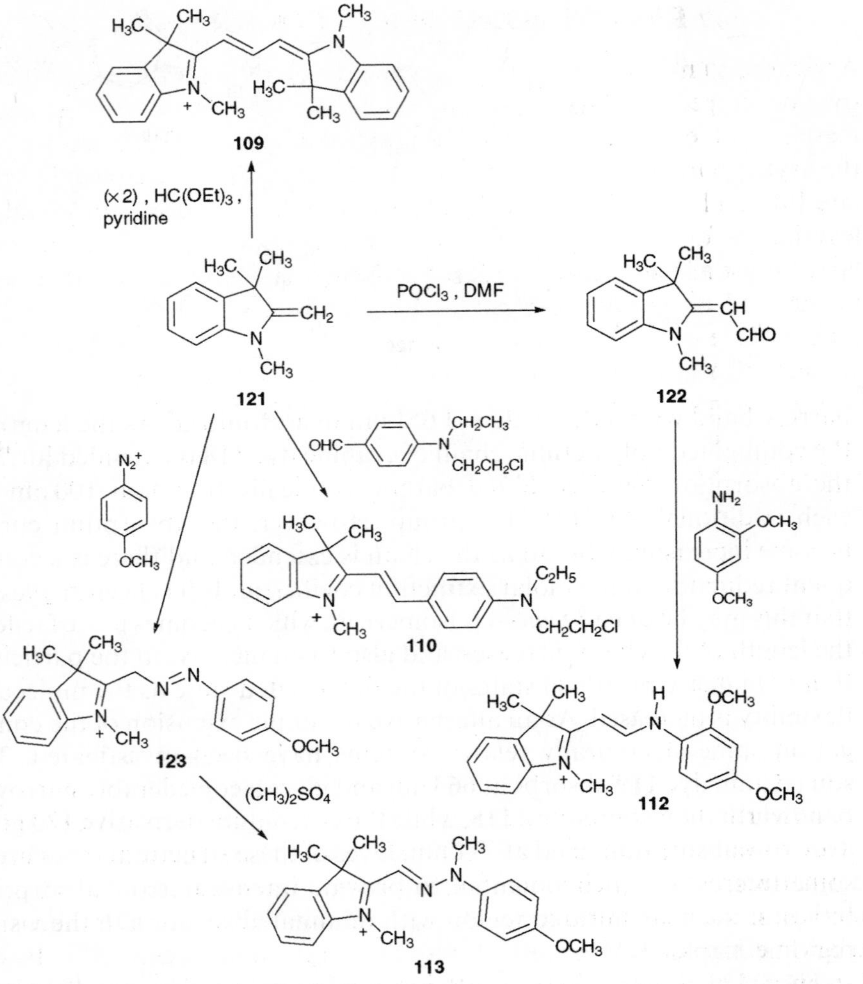

Scheme 6.1   *Synthetic approach to some indolenine-based polymethine dyes*

heating Fischer's base with an aldehyde, as illustrated for the case of C. I. Basic Violet 7, **110**. The azacyanine C. I. Basic Yellow 11 (**112**) is synthesised by the condensation reaction of Fischer's aldehyde **122** with 2,4-dimethoxyaniline. In the synthesis of diazacyanine dye C. I. Basic Yellow 28 (**113**), an important golden yellow dye for acrylic fibres, Fischer's base **121** is treated with 4-methoxybenzenediazonium chloride and undergoes an azo coupling reaction at the methylene group to give the azo dye **123**. Methylation of this dye with dimethyl sulfate gives the diazacyanine dye **113**.

ARYLCARBONIUM ION COLORANTS

Arylcarbonium ion colorants were historically the first group of synthetic dyes developed for textile applications. In fact, Mauveine, the first commercial synthetic dye, belonged to this group (Chapter 1). The majority of the arylcarbonium ion colorants still in use today were discovered in the late 19th and early 20th centuries. As a group they are used considerably less than in former times, but many are still of some importance, particularly for use as basic (cationic) dyes for the coloration of acrylic fibres and paper, and as pigments. Structurally, they are closely related to the polymethine dyes, especially the cyanine types, and they tend to show similar properties. For example, they provide extremely intense, bright colours, covering virtually the complete shade range, but they are generally inferior in technical properties compared with the azo, carbonyl and phthalocyanine chemical classes and as a result their importance has declined over the years.

124   125

Arylcarbonium ion dyes encompass a diversity of structural types. Most of the dyes are cationic but there are some neutral and anionic derivatives. The best-known arylcarbonium ion dyes are the diarylmethines, such as Auramine O, C. I. Basic Yellow 2 (**124**) and the triarylmethines, the simplest of which is Malachite Green, C. I. Basic Green 4, **125**. The essential structural feature of these two groups is a central carbon atom attached to either two or three aromatic rings. Of these two groups, the triarylmethines are generally the most stable and thus the most useful. Commonly, they are referred to as triaryl*methanes* (or triphenylmethanes), but the name triaryl*methines* more correctly indicates that the carbon atom to which the aromatic rings are attached is sp^2, rather than sp^3, hybridised. Aza analogues of these dyes in which the central carbon is replaced by a nitrogen atom are also conveniently included in this class. There are also a number of derivatives obtained by bridging the di- and triarylmethines and their aza analogues across the *ortho–ortho'* positions of two of the aromatic rings with a heteroatom. Examples of these types of heterocyclic systems, which may be represented by the general structure **126**, are illustrated in Table 6.1.

Table 6.1 *Heterocyclic arylcarbonium ion dyes and their aza analogues*

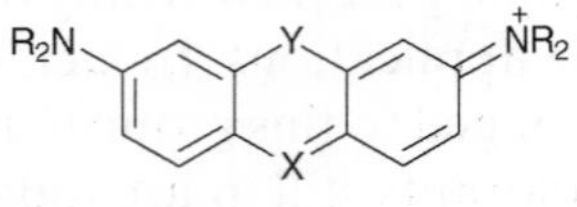

126

X	Y	Type
–C(Ar)=	–O–	Xanthene
–C(Ar)=	–S–	Thioxanthene
–C(Ar)=	–NR–	Acridine
–N=	–O–	Oxazine
–N=	–S–	Thiazine
–N=	–NR–	Azine

Mauveine, the original synthetic dye, was of the azine type, its principal component being compound **127**. This particular group of dyes are now essentially only of historic interest. Xanthene dyes, such as Rhodamine B, C. I. Basic Violet 10 (**128**) and fluorescein (**129**), are relatively inexpensive dyes which are notable for their strong fluorescence, a feature which has been attributed to their structural rigidity. They are used in a wide range of applications, including the detection of defects in engineered articles and in the tracking of water currents. Rhodamines are also used in dye lasers (Chapter 10)

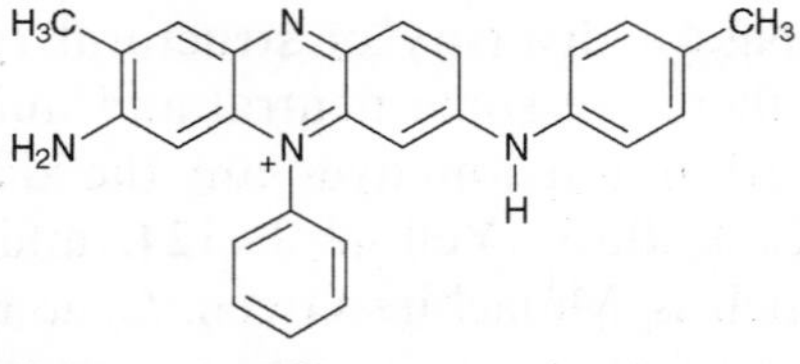

127

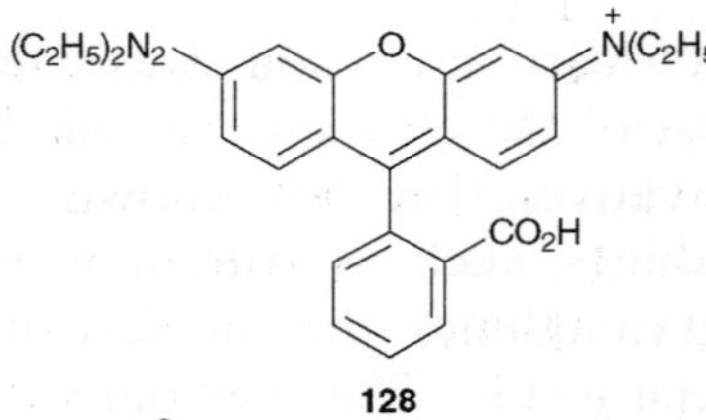

128

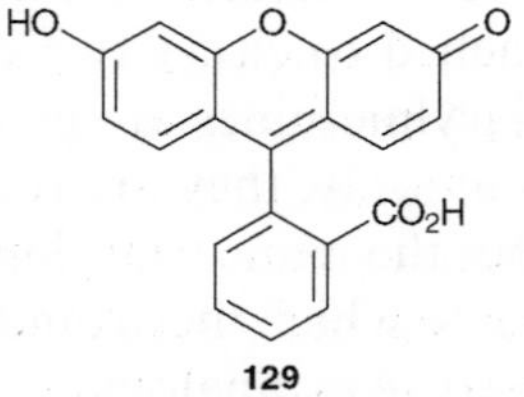

129

The valence-bond (resonance) description of the triphenylmethine dye Malachite Green (**125**) is illustrated in Figure 6.5. Comparison with Figure 6.4 reveals their structural similarity compared with cyanine dyes. Formally, the dye contains a carbonium ion centre, as a result of a contribution from resonance form II. The molecule is stabilised by resonance that involves delocalisation of the positive charge on to the *p*-amino

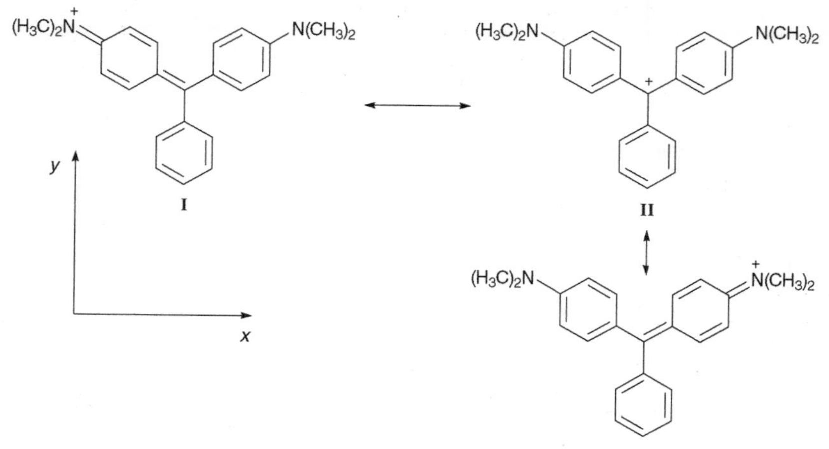

Figure 6.5   *Valence-bond (resonance) approach to Malachite Green* **125**

nitrogen atoms as illustrated by forms I and III. Because of the steric constraints imposed by the presence of the three rings, triarylmethine dyes cannot adopt a planar conformation. The three rings are twisted out of the molecular plane, adopting a shape like a three-bladed propeller. Malachite Green shows two absorption bands in the visible region with λ_{max} values of 621 and 428 nm. Hence, its observed green colour is due to the addition of blue and yellow components. The long wavelength band is polarised along the x-axis and the short wavelength band along the y-axis.

The synthesis of arylcarbonium ion dyes and pigments generally follow a similar set of principles. A few selected examples are shown in Scheme 6.2 to illustrate these principles. Essentially, the molecules are constructed from aromatic substitution reactions. In general, a C_1 electrophile, for example phosgene ($COCl_2$), formaldehyde, chloroform or carbon tetrachloride, reacts with an aromatic system which is activated to electrophilic attack by the presence of a strongly electron-releasing group such as the amino (primary, secondary or tertiary) or hydroxy group. Symmetrical diarylmethines and triarylmethines may be synthesised in one operation, for example by reaction of one mole of phosgene with either two or three moles of the appropriate aromatic compound. Depending on the particular electrophile used, an oxidation may be required at some point in the reaction sequence to generate the final product. Two methods of synthesis of the diarylmethine Auramine O (**124**) are shown in Scheme 6.2. Formerly, this yellow dye was prepared from Mischler's Ketone (**130**), which may be obtained from the reaction of dimethylani-

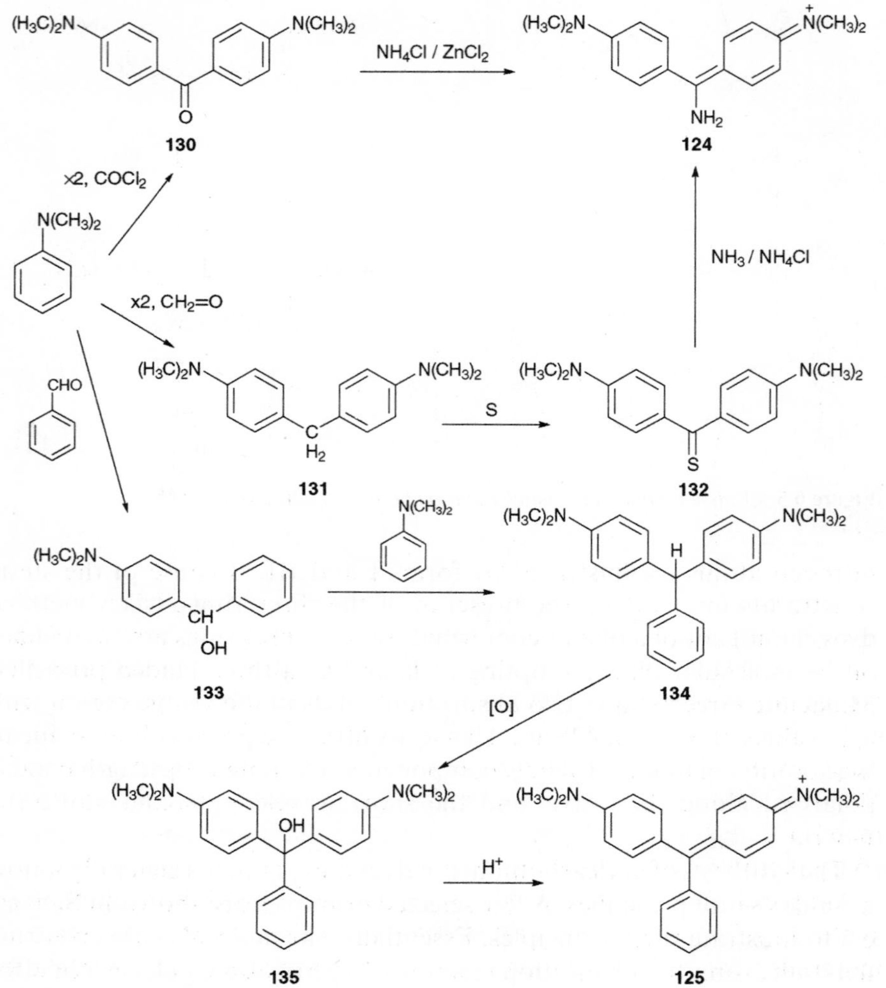

Scheme 6.2   *Synthetic approach to some arylcarbonium ion dyes*

line (2 mol) with phosgene (1 mol). In the currently preferred method, dimethylaniline (2 mol) is reacted with formaldehyde (1 mol) to give the diaryl compound **131**. This compound is then heated with sulfur and ammonium chloride in a stream of ammonia at 200 °C. The dye **124** is formed *via* the thiobenzophenone **132** as an intermediate. The synthesis of Malachite Green, **125**, is given in Scheme 6.2 to illustrate how an unsymmetrical triarylmethine derivative may be prepared. Dimethylaniline is reacted with benzaldehyde under acidic conditions to give the alcohol **133**. This compound is then treated with a further equivalent of

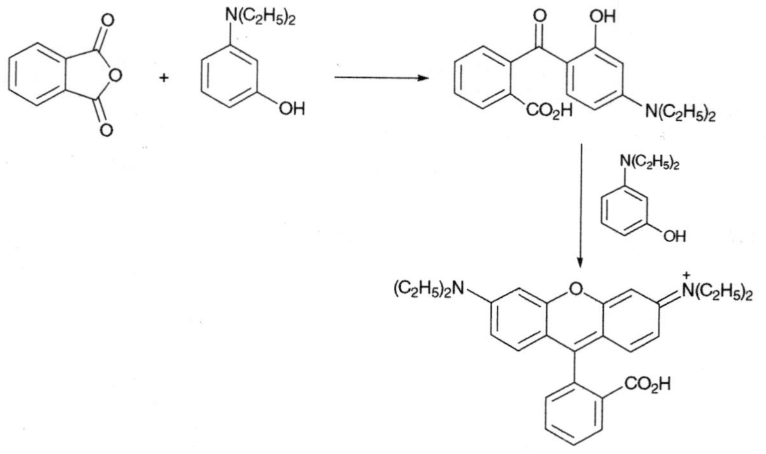

Scheme 6.3 *Synthesis of Rhodamine B,* **128**

dimethylaniline to give the leuco base **134**, which is subsequently oxidised to the carbinol base **135**. Acidification of this compound leads to Malachite Green (**125**). For many years, lead dioxide (PbO_2) was the agent of choice for oxidation reactions of this type. Lead-free processes, for example using air or chloranil in the presence of various transition metal catalysts, are now preferred for toxicological and environmental reasons. By using different aromatic aldehydes and aromatic amines as starting materials, this method may be adapted to produce a wide range of triarylmethine dyes.

As an example of a heterocyclic arylcarbonium ion dye, the method of synthesis of Rhodamine B (**128**) is shown in Scheme 6.3. The starting materials in this case are phthalic anhydride and 3-*N*,*N*-diethylaminophenol.

Each of the products whose synthesis is illustrated in Schemes 6.2 and 6.3 are coloured cationic species. When the counter-anion is chloride, the products are water-soluble and useful as basic dyes for acrylic fibres. Precipitation of cationic dyes of these types from aqueous solution using large polymeric counter-anions, notably phosphomolybdates, phosphotungstates and phosphomolybdotungstates leads to a range of highly insoluble red, violet, blue and green pigments. These pigments exhibit high brilliance and intensity of colour and high transparency, and are thus well suited to some printing ink applications.

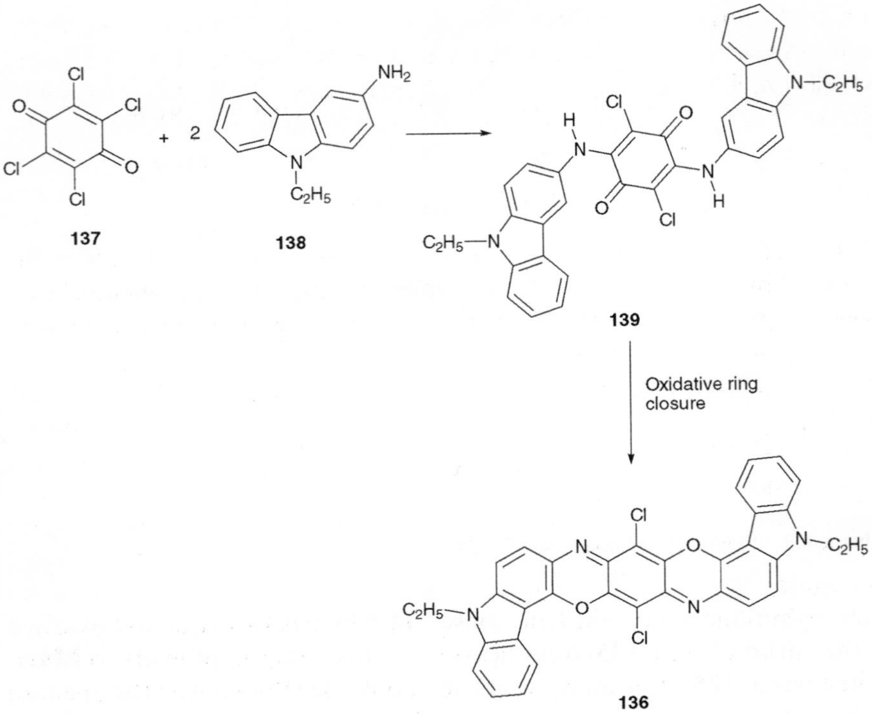

Scheme 6.4 *Synthesis of Carbazole Violet,* **136**

DIOXAZINES

Dioxazine colorants, as the name implies, contain two oxazine ring systems as the chromophoric grouping. They are relatively few in number and generally restricted to violet to blue shades. Probably the most important dioxazine colorant is C. I. Pigment Violet 23, **136**. This product is usually referred to as Carbazole Violet and is the most significant violet pigment for high-performance applications (Chapter 9). Compound **136** has been shown to have an angular structure as illustrated in Scheme 6.4 rather than the linear structure that is shown in most older texts. The pigment is characterised by a brilliant intense reddish-violet colour, very good lightfastness and resistance to heat and solvents. Its synthesis is illustrated in Scheme 6.4.

In the synthetic scheme, 3-amino-9-ethylcarbazole **138** (2 mol) is condensed with chloranil **137** (1 mol) to form the intermediate **139**. This intermediate is then converted into the dioxazine pigment **136** by oxidative cyclisation at around 180 °C in an aromatic solvent and in the presence of a catalyst such as aluminium(III) chloride or benzenesulfonyl chloride. Sulfonation of this pigment gives rise to water-soluble dyes which may be used as direct dyes for cotton. An example is C. I. Direct Blue 108 which contains 3 or 4 sulfonic acid groups. The dioxazine system is also used as the chromophoric group in some reactive dyes (Chapter 8).

SULFUR DYES

Sulfur dyes are a group of low cost dyes used in the coloration of cellulosic fibres. The dyes are fairly small in number although some of the individual products are manufactured in very large quantities. They are capable of providing a wide range of hues although they tend to give rise to rather dull colours, and they are of particular importance as blacks, navy blues, browns and olive greens. In spite of the fact that these products have been known for many years, the chemical structures of sulfur dyes are by no means completely established. They are known to be complex mixtures of polymeric molecular species containing a large proportion of sulfur in the form of sulfide (–S–), disulfide (–S–S–) and polysulfide ($-S_n-$) links and in heterocyclic rings. C. I. Sulfur Black 1 is by far the most important product in the series. In fact, it may well be the individual dye of all chemical types that has the largest production volume worldwide. It has been suggested that some sulfur dyes are based on the phenothiazonethioanthrone chromophoric system shown in Figure 6.6.

Traditional sulfur dyes are products of high insolubility in water. They are applied to the cellulosic fibres after conversion into a water-soluble leuco form by treatment with an aqueous alkaline solution of sodium sulfide. The chemistry of this process is thought to involve mainly the reduction by the sulfide anion of disulfide (–S–S–) linkages, leading to alkali-soluble thiol (–SH) groups. After application of the leuco form to the fibre, the insoluble polymeric structure of the dye is regenerated by air oxidation and becomes trapped within the fibre. A second group of sulfur dyes are pre-formed leuco dyes which are 'ready-to-use' in concentrated aqueous solutions. A third group contains thiosulfate ($-S-SO_3^-$ Na^+) water-solubilising groups. These dyes, referred to as Bunte salts, are applied to the fibres together with sodium sulfide. During their application the thiosulfate groups are reduced to form dimeric and polymeric species as a result of disulfide bond formation. The particular advantage of sulfur dyes, as a class of dyes for cellulosic fibres, is that they provide reasonable technical performance at low cost. Unfortunately sulfur dyes present some environmental problems, largely associated with sulfide residues which remain in the dyehouse effluent and this feature may cause their use to decline in years to come.

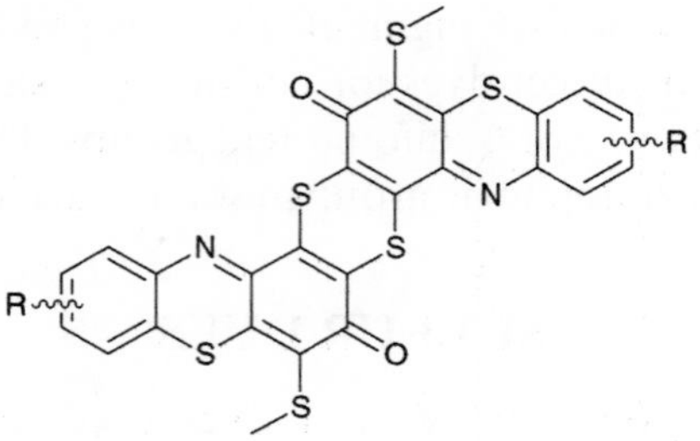

Figure 6.6   *The phenothiazonethioanthrone chromophoric system proposed as a constituent of some sulfur dyes*

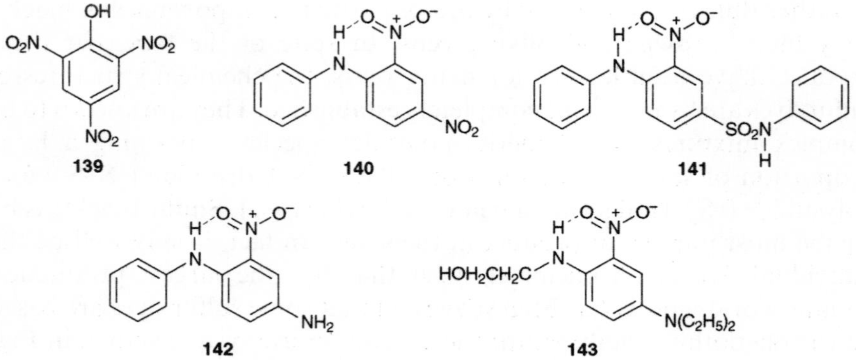

Figure 6.7   *Structure of some nitro dyes*

The manufacture of sulfur dyes involves *sulfurisation* processes, the chemistry of which remains rather mysterious and may arguably be considered still to be in the realms of alchemy! The processes involve heating elemental sulfur or sodium polysulfide, or both, with aromatic amines, phenols or aminophenols. These reactions may be carried out either as a dry bake process at temperatures between 180 and 350 °C or in solvents such as water or aliphatic alcohols at reflux or at even higher temperatures under pressure. C. I. Sulphur Black 1, for example, is prepared by heating 2,4-dinitrophenol with sodium polysulfide.

NITRO DYES

The nitro group is commonly encountered as a substituent in dyes and pigments of most chemical classes, but it acts as the essential chromophore in only a few dyes. Nitro dyes are a small group of dyes of some importance as disperse dyes for polyester and as semi-permanent hair dyes. Picric acid, **139**, was historically the first nitro dye, although it was never really commercially important due to its poor dyeing properties, its toxicity and its potential explosive properties (Chapter 1). The nitro dyes used today have relatively simple nitrodiphenylamine structures. Some examples of nitro dyes are shown in Figure 6.7. They contain at least one nitro (NO_2) group as the chromophore and electron acceptor and the electron-releasing amino group completes the donor–acceptor system. Nitro dyes are capable of providing bright yellow, orange and red shades, but the colours are amongst the weakest provided by the common commercial chromophores. Nitrodiphenylamines are nevertheless of some importance as yellow disperse dyes, such as C. I. Disperse Yellow 42 (**141**), because of their low cost and their good lightfastness. The good lightfastness of these dyes is attributed to the intramolecular hydrogen bonding between the *o*-nitro group and the amino group while the electron-withdrawing (nitro or sulfonamide) group in the *para*-position is also important as it gives rise to an increase in tinctorial strength. A wider range of hues is provided by some nitro hair dyes, such as compounds **142**, which is red, and **143** which is violet.

The synthesis of nitro dyes is relatively simple, a feature which accounts to a certain extent for their low cost. The synthesis, illustrated in Scheme 6.5 for compounds **140** and **141**, generally involves a nucleophilic substitution reaction between an aromatic amine and a chloronitroaromatic compound. The synthesis of C. I. Disperse Yellow 14 (**140**) involves the reaction of aniline with 1-chloro-2,4-dinitroaniline while compound **141** is prepared by reacting aniline (2 mol) with compound **144** (1 mol).

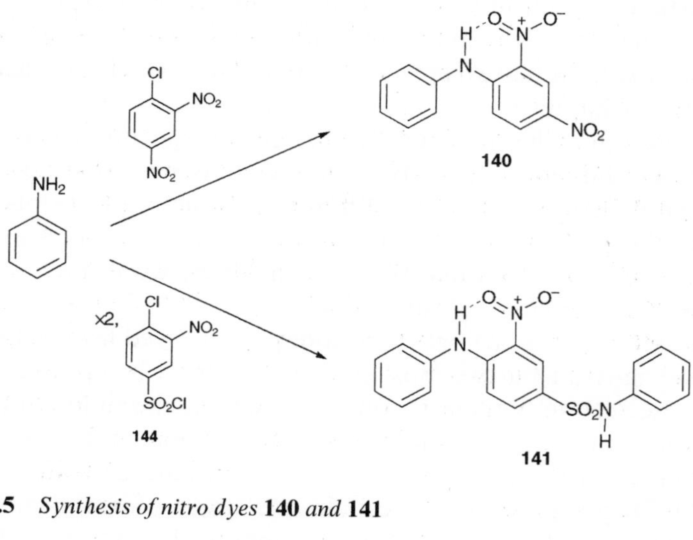

Scheme 6.5 *Synthesis of nitro dyes* **140** *and* **141**

Chapter 7

Textile Dyes (Excluding Reactive Dyes)

In Chapters 3–6, the commercially important chemical classes of dyes and pigments are discussed in terms of their essential structural features and the principles of their synthesis. The reader will encounter further examples of these individual chemical classes of colorants throughout Chapters 7–10 which, as a complement to the content of the earlier chapters, deal with the chemistry of their application. Chapters 7, 8 and 10 are concerned essentially with the application of dyes, whereas Chapter 9 is devoted to pigments. The distinction between these two types of colorants has been made previously in Chapter 2. Dyes are used in the coloration of a wide range of substrates, including paper, leather and plastics, but by far their most important outlet is on textiles. Textile materials are used in a wide variety of products, including clothing of all types, curtains, upholstery and carpets. This chapter deals with the chemical principles of the main application classes of dyes that may be applied to textile fibres, except for reactive dyes, which are dealt with exclusively in Chapter 8.

Textile fibres may be classified into three broad groups: natural, semi-synthetic and synthetic. Unlike dyes and pigments, which are now almost entirely synthetic in origin, natural fibres continue to play a prominent part in textile applications. The most important natural fibres are either of animal origin, for example the protein fibres, wool and silk, or of vegetable origin, such as cotton, which is a cellulosic fibre. The only significant semi-synthetic fibres used today are derived from cellulose as the starting material, the two most important of these being viscose rayon and cellulose acetate. Viscose rayon is a regenerated cellulosic fibre. It is manufactured by reacting cellulose, *e.g.* from wood pulp, with carbon disulfide in alkali to give its water-soluble xanthate derivative. This is followed by regeneration of the cellulose in fibrous form using sulfuric acid. Cellulose acetate is a chemically modified cellulose derivative, manufactured by acetylation and partial hydrolysis of cotton. The most important completely synthetic fibres are polyester, polyamides (nylon), and acrylic fibres. Textile fibres share the common feature that they are made up of polymeric organic molecules, but the physical and chemical nature of the polymers involved vary widely and this explains why each type of fibre essentially requires its own 'tailor-made' application classes of dyes.

Dye molecules are designed to ensure that they have a set of properties that are appropriate to their particular applications. The most obvious requirement for a dye is that it must possess the desired colour, in terms of hue, strength and brightness. The relationships between colour and constitution of dyes has been discussed principally in Chapter 2, although the reader will find specific aspects relating to particular chemical classes in Chapters 3–6. A further feature of dye molecules, which is of some practical importance, is their ability to dissolve in water. Since textile dyes are almost always applied from an aqueous dyebath solution, they are required to be either soluble in water or, alternatively, to be capable of conversion into a water-soluble form. Many dye application classes, including acid, mordant, premetallised, direct, reactive and cationic dyes, are readily water-soluble. Disperse dyes for polyester, in contrast, are only sparingly soluble in water, but they have sufficient solubility for their application at high temperatures. A few groups of dyes, including vat and sulfur dyes for cellulosic fibres, are initially insoluble in water and are thus essentially pigments. However, they may be converted chemically into a water-soluble form and in this form they can be applied to the fibre, after which the process reversed and the insoluble form is regenerated in the fibre.

Dyes must be firmly attached to the textile fibres to which they are applied in order to resist removal, for example by washing. This may be achieved in a number of ways. The molecules of many dye application classes are designed to provide forces of attraction for the polymer molecules which constitute the fibre. In the case of reactive dyeing, the dye molecules combine chemically with the polymer molecules forming covalent bonds (Chapter 8). In further cases, for example vat, sulfur and azoic dyes for cellulosic fibres, an insoluble pigment is generated within the fibres and is retained by mechanical entrapment. In other cases, a set of dye–fibre intermolecular forces operate which, depending on the particular dye–fibre system, is commonly a combination of ionic, dipolar, van der Waals' forces and hydrogen bonding. An additional feature of textile dyeing is that the dye must distribute itself evenly throughout the material to give a uniform colour, referred to as a level dyeing. Finally, the dye must provide an appropriate range of fastness properties, for example to light, washing, heat, *etc.* This chapter provides an overview of the most important application classes of dyes for textiles, with a range of selected examples that illustrate how the dye molecules are designed to suit their particular application. The discussion is organised according to fibre type in three sections: dyes for protein fibres, dyes for cellulosic fibres and dyes for synthetic fibres. In each case there is a description of the structure of the polymer, followed by a discussion of the structural features of the dyes that determine their suitability for application to the particular type of fibre. The subject of reactive dyes is of sufficient interest and importance to warrant separate treatment (Chapter 8).

DYES FOR PROTEIN FIBRES

Protein fibres are natural fibres derived from animal sources, the most important of these being wool and silk. The principal component of the wool fibre is the protein keratin, the molecular structure of which is illustrated in outline in Figure 7.1. The protein molecules consist of a long polypeptide chain constructed from the eighteen commonly encountered amino acids that are found in most naturally-occurring proteins. The structures of these amino acids are well documented in general chemical and biochemical textbooks and so they are not reproduced here. As a result of the diverse chemical nature of these amino acids, the protein side-chains (R^1, R^2, R^3 in Figure 7.1) are of widely varying character, containing functionality which includes, for example, amino and imino, hydroxy, carboxylic acid, thiol and alkyl groups and heterocyclic functionality. At intervals, the polypeptide chains are linked together by disulfide (–S–S–) bridges derived from the amino acid cystine. There are also ionic links between the protonated amino (–NH_3^+) and carboxylate (–CO_2^-) groups, which are located on the amino acid side-groups and at the end of the polypeptide chains. Many of the functional groups on the wool fibre play some part in the forces of attraction involved when dyes are applied to the fibres. Protein fibres may be dyed using a number of application classes of dyes, the most important of which are acid, mordant and premetallised dyes, the structural features of which are discussed in the rest of this section, and reactive dyes which are considered separately in Chapter 8.

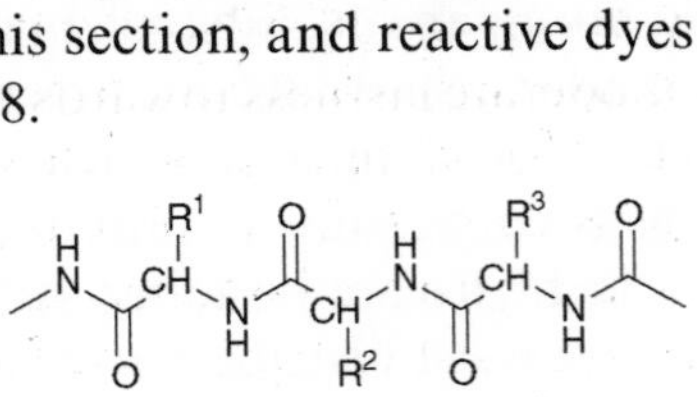

Figure 7.1   *Structure of the protein keratin*

Acid dyes derive their name historically from the fact that they are applied to protein fibres such as wool under acidic conditions. They are also used to a certain extent to dye polyamide fibres such as nylon. Acid dyes may be conveniently classified as either acid-leveling or acid-milling types. Acid-leveling dyes are group of dyes that show only moderate affinity for the wool fibres. Because the intermolecular forces between the dye and the fibre molecules are not strong, these dyes are capable of migrating through the fibre and thus produce a level dyeing. Acid-milling dyes are a group of dyes, which show much stronger affinity for the wool fibres. Because of the strength of the intermolecular forces between the dye and the fibre molecules, the dyes are less capable of migration and this can present difficulties in producing level dyeings. However, they give superior fastness to washing.

A characteristic feature of acid dyes for protein and polyamide fibres is the presence of one or more sulfonate ($-SO_3^-$) groups, usually as sodium (Na^+) salts. These groups have a dual role. Firstly, they provide solubility in water, the medium from which the dyes are applied to the fibre. Secondly, they ensure that the dyes carry a negative charge (*i.e.* they are anionic). When acid conditions are used in the dyeing process, the protein molecules acquire a positive charge. This is due mainly to protonation of the amino ($-NH_2$) and imino ($=NH$) groups on the amino acid side-chains, to give $-NH_3^+$ and $=NH_2^+$ groups respectively, and to the suppression of the ionisation of the carboxylic acid groups. The positive charge on the polymer attracts the acid dye anions by ionic forces, and these displace the counter-anions within the fibre by an ion exchange process. As well as these ionic forces of attraction, van der Waals' forces, dipolar forces and hydrogen bonding between appropriate functionality of the dye and fibre molecules may also play a part in the acid-dyeing of protein fibres. In terms of size and shape, often an important consideration in the design of dye molecules, acid-leveling dyes may be described as small to medium-sized planar molecules. This allows the dyes to penetrate easily into the fibre and also permits a degree of movement or migration within the fibre as the ionic bonds between the dye and the fibre are capable of breaking and then re-forming, thus producing a level or uniform colour. However, as the dye is not very strongly bonded to the fibre, it may show only moderate fastness towards wet-treatments such as washing. Acid-milling dyes are significantly larger molecules than acid-leveling dyes and they show enhanced affinity for the fibre, and hence improved fastness to washing, as a result of more extensive van der Waals' forces, dipolar forces and hydrogen bonding.

Most acid dyes, especially yellows, oranges and reds, belong to the azo chemical class while blues and greens are often provided by carbonyl

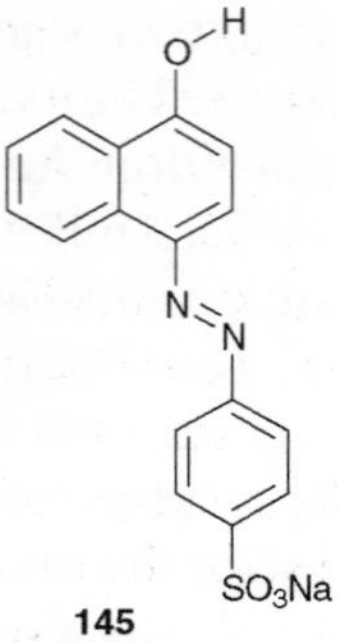

145

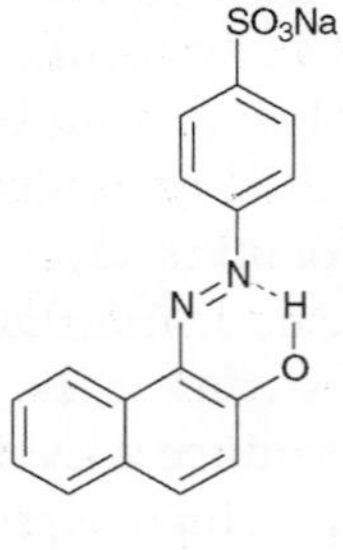

146

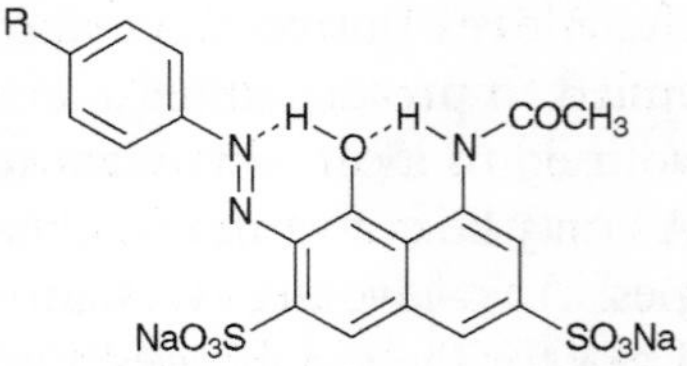

147a R = H; **147b** R = $C_{12}H_{25}$

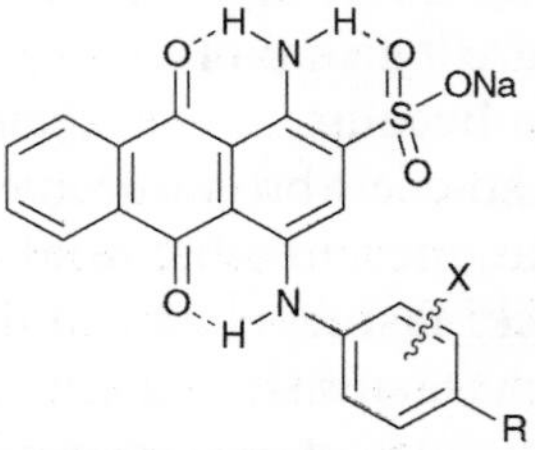

149a R = H, X = H;
149b R = $C_{12}H_{25}$, X = SO_3Na

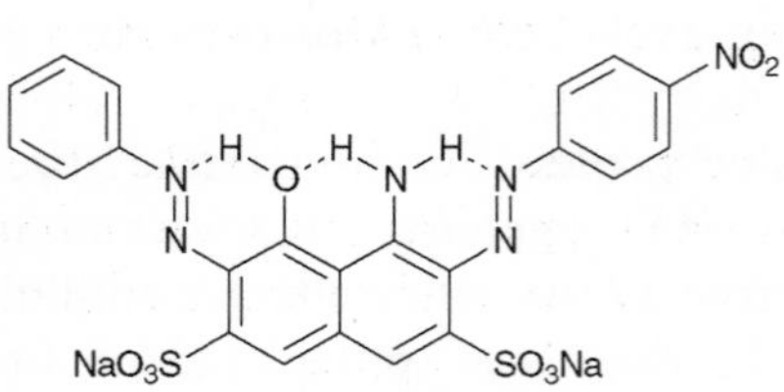

148

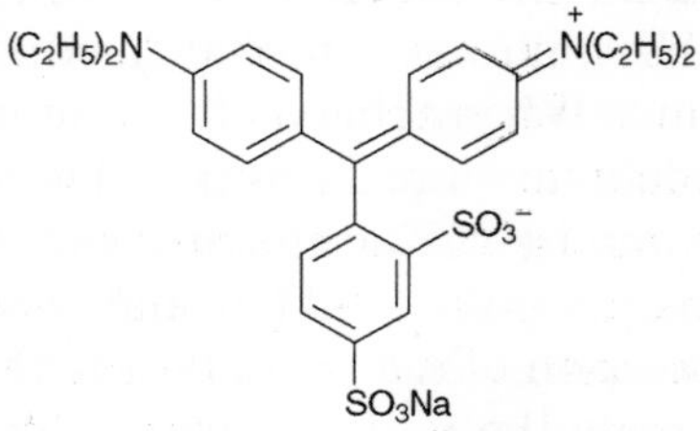

150

Figure 7.2 *Structures of some typical acid dyes for protein fibres*

dyes, especially anthraquinones, and to a certain extent by arylcarbonium ion types. Figure 7.2 illustrates some typical acid dye structures. A notable aspect of the structure of dyes **146–149** is the strong intramolecular hydrogen-bonding which exists in the form of six-membered rings, a feature which enhances the stability of the compounds and, in particular, confers good lightfastness properties. This has been explained by a reduction in electron density at the chromophore as a result of the hydrogen-bonding, reducing the sensitivity of the dye towards photochemical oxidation. For this reason, intramolecular hydrogen-bonding is a feature commonly encountered in the structures of a wide range of dyes and pigments. Intramolecular hydrogen-bonding also reduces the acidity of a hydroxy group, and thus can lead to improved resistance towards alkali treatments. A comparison between the two isomeric monoazo acid dyes C. I. Acid Orange 20 (**145**) and C. I. Acid Orange 7 (**146**) illustrates the effect of intramolecular hydrogen bonding. Dye **146** shows significantly improved fastness to alkaline washing and lightfastness compared with dye **145** in which intramolecular hydrogen bonding is not possible. A comparison of the structurally related monoazo dyes **147a** (C. I. Acid Red 1) and **147b** (C. I. Acid Red 138), and of the anthraquinone acid dyes **149a** (C. I. Acid Blue 25) and **149b** (C. I. Acid Blue 138) illustrates the distinction between acid leveling and acid-milling dyes. Dyes **147b** and **149b** show excellent resistance to washing as a result of the presence of the long alkyl chain substituent ($C_{12}H_{25}$), which is attracted to hydrophobic or non-polar parts of the protein fibre molecules by van der Waals' forces. Because of the extremely strong dye–fibre affinity, dyes of this type are often referred to as acid supermilling dyes. Dyes **147a**, **149a** (C. I. Acid Black 1), **148**, a typical disazo acid dye, and C. I. Acid Blue 1 (**150**), an example of a triphenylmethine acid dye, are acid-leveling dyes. In the case of dye **150**, note that while the nitrogen atoms carry a formal single delocalised positive charge, the presence of two sulfonate groups ensures that the dye overall is anionic.

The ability of transition metal ions, and especially chromium (as Cr^{3+}), to form very stable metal complexes may be used to produce dyeings on protein fibres with superior fastness properties, especially towards washing and light. The chemistry of transition metal complex formation with azo dyes is discussed in some detail in Chapter 3. There are two application classes of dyes in which this feature is utilised, mordant dyes and premetallised dyes, which differ significantly in application technology but involve similar chemistry.

Mordant dyes generally have the characteristics of acid dyes but with the ability in addition to form a stable complex with chromium. Most commonly, this takes the form of two hydroxy groups on either side of (*ortho* to) the azo group of a monoazo dye, as illustrated for the case of C. I. Mordant Black 1 (**151**). The dye is generally applied to the fibre as an acid dye and then treated with a source of chromium, commonly sodium or potassium dichromate. As a result of the process, the chromium(VI) is reduced by functional groups on the wool fibre, for example the cysteine thiol groups, and a chromium(III) complex of the dye is formed within the

151

fibre by a process such as that illustrated in Figure 7.3. A dye of this type acts as a tridentate ligand, the chromium bonding with two oxygen atoms derived from the hydroxy groups and with one nitrogen atom of the azo group. The complexes formed are six-coordinate with octahedral geometry. It has not been established with certainty how the remaining three valencies of chromium are satisfied in the mordant dyeing of protein fibres. There are a number of possibilities, which include bonding with water molecules, with coordinating groups ($-OH$, $-SH$, $-NH_2$, $-CO_2H$, etc.) on the amino acid side-chains on the fibre, or with another dye molecule. The principal problem currently with the use of chrome mordant dyes is environmental, associated mainly with the presence of residual chromium, an undesirable heavy metal, in dyehouse effluent.

Premetallised dyes, as the name implies, are pre-formed metal complex dyes. They are usually six-coordinate complexes of chromium(III) with octahedral geometry, as exhibited for example by C. I. Acid Violet 78, **152**, although some complexes of cobalt(III) are also used. Most premetallised dyes are azo dyes, with one nitrogen of the azo group playing a part in complexing with the central metal ion. Since in this case there are two azo dye molecules coordinated with one chromium atom, compound **152** is referred to as a 2:1 complex. 1:1 Complexes are also used, but to a lesser extent. Premetallised dyes of this type, like traditional acid dyes, are anionic in nature even though, as is the case with compound **152**, they may not contain sulfonate groups. Indeed, the presence of sulfonate groups can cause the dye anions to be too strongly attracted to the fibre, which leads in turn to levelness problems. The purpose of the sulfone group in dye **152** is to enhance the hydrophilic character of the molecule and hence its water solubility, without increasing the charge on the dye anion. Premetallised dyes are applied to protein fibres as acid dyes and, because of the special stability of chromium(III) complexes due to their d^3 configuration, provide dyeings with excellent fastness properties.

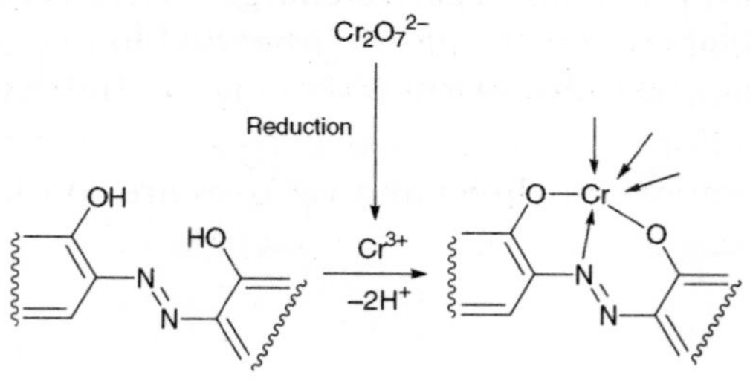

Figure 7.3   *Chemistry of chrome mordanting*

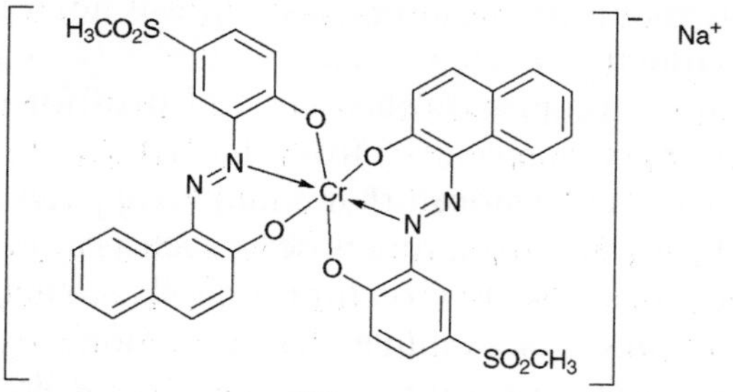

152

DYES FOR CELLULOSIC FIBRES

Cellulosic fibres are natural fibres derived from plant sources. The most important cellulosic fibres are cotton, viscose, linen, jute, hemp and flax. The principal component of the cotton fibre is cellulose, the structure of which is shown in Figure 7.4. Cotton is in fact almost pure cellulose (up to 95%). Cellulose is a polysaccharide. It is a high molecular weight polymer consisting of long chains of repeating glucose units, with up to around 1300 such units in each molecule. Cellulose has a fairly open structure, which allows large dye molecules to penetrate relatively easily into the fibre. Each glucose unit contains three hydroxy groups, two of which are secondary and one primary, and these give the cellulose molecule a considerable degree of polar character. The presence of the hydroxy groups is of considerable importance in the dyeing of cotton. For example, the ability of the hydroxy groups to form intermolecular hydrogen bonds is thought to be of some importance in direct dyeing, while reactive dyeing (Chapter 8) involves a chemical reaction of the hydroxy groups with the dye to form dye–fibre covalent bonds. The tendency of the hydroxy groups to ionise to a certain extent (to $-O^-$) means that the fibres can carry a small negative charge. There are a larger number of application classes of dyes that may be used to dye cellulosic fibres such as cotton than for any other fibre. These application classes include direct, vat, sulfur, azoic and reactive dyes. The structural features of two of the most important, direct and vat dyes are considered in this section while the discussion of reactive dyes is continued separately in Chapter 8.

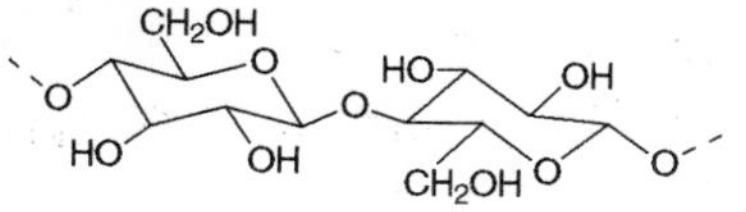

Figure 7.4 *Structure of cellulose*

An outline of the chemistry of sulfur dyes, which is not well established, is presented in Chapter 6.

Direct dyes are a long-established class of dyes for cellulosic fibres. They derive their name historically from the fact that they were the first application class to be developed that could be applied directly to these fibres without the need for a fixation process such as mordanting. In some ways, direct dye molecules are structurally similar to acid dye molecules used for protein fibres. For example, they are anionic dyes as a result of the presence of sulfonate $(-SO_3^-)$ groups. However, the role of the sulfonate groups in the case of direct dyes is simply to provide water solubility. In contrast to the acid dyeing of protein fibres, ionic attraction to the fibre is not involved in the direct dyeing of cellulosic fibres. In fact, the anionic nature of the dyes can reduce affinity for the fibres because cellulosic fibres may carry a small negative charge. For this reason, usually only as many sulfonate groups as are required to give adequate solubility in water are present in direct dyes and, in addition, the groups are distributed evenly throughout the molecules. Arguably the most important features of direct dye molecules which influence their application properties are their size and shape. They are, in general, large molecules and in shape they are long, narrow and planar. Direct dyes show affinity for cellulose by a combination of van der Waals', dipolar and hydrogen-bonding intermolecular forces. Individually, these forces are rather weak. The long, thin and flat geometry of the molecules is essential to allow the dye molecules to align with the long polymeric cellulose fibre molecules and hence to maximise the overall effect of the combined set of intermolecular forces. Chemically, direct dyes, of which compound **153** (C. I. Direct Orange 25) is a typical example, are almost invariably azo dyes, commonly containing two or more azo groups. The long, flat, linear shape of compound **153** allows groups such as the –OH, –NHCO (amide), and –N=N– groups in principle to form hydrogen bonds with OH groups on cellulose as it lines up with the cellulose molecule. There are only two sulfonate groups in **153** and these are well separated. This is sufficient to give adequate water solubility for their application. Also, it may be argued that the sulfonate groups are on the opposite side of the molecule from groups which may be participating in

153

hydrogen bonding with the fibre and this means that they will be oriented away from the cellulose molecule, thus minimising any negative charge repulsion effects

Direct dyes, in comparison particularly with vat dyes and reactive dyes, provide only moderate washfastness. They are, however, inexpensive and are therefore the dyes of choice for applications, such as the coloration of paper, where cost is of prime concern and fastness to wet treatments is of lesser importance. For textile applications, the washfastness may in certain cases be improved by certain chemical aftertreatments. For example, a group of products referred to as *direct and developed* dyes contain free amino ($-NH_2$) groups. Treatment of the dyed fabric with aqueous sodium nitrite under acidic conditions results in diazotisation of these groups and the resulting diazonium salts may be reacted with a variety of coupling components, such as 2-naphthol. The larger azo dye molecule, which is thus formed, is more strongly attracted to the fibre and less soluble in water, both features leading to improved washfastness. Alternatively, in a process referred to as *after-coppering*, some *o,o'*-dihydroxyazo direct dyes may be treated with copper(II) salts to form square planar metal complexes which also show improved washfastness properties.

Vat dyes are a group of totally water-insoluble dyes and in this respect they are essentially pigments. In fact, some vat dyes, after conversion into a suitable physical form, may be used as pigments (Chapters 4 and 9). For application to cellulosic fibres, vat dyes are initially converted into a water-soluble 'leuco' form by an alkaline reduction process. Commonly, this conversion is carried out using a reducing agent such as sodium dithionite ($Na_2S_2O_4$) in the presence of sodium hydroxide. In the leuco form, the dyes are taken up by the cellulosic fibre. Subsequently, they are converted back by an oxidation reaction to the insoluble pigment. Hydrogen peroxide is commonly used as the oxidising agent for this process although atmospheric oxygen can also effect the oxidation under appropriate conditions. The pigment or molecular aggregate thus produced becomes trapped mechanically within the fibre and its insolubility gives rise to the excellent washfastness properties which are characteristic of vat dyes. The vat dyeing process is usually completed with a high temperature aqueous surfactant treatment, or 'soaping', which enhances the molecular aggregation process and develops crystallinity. The chemistry of the reversible reduction–oxidation process involved in vat dyeing, which is illustrated in Figure 7.5, requires the presence in the vat dye of two carbonyl groups linked *via* a conjugated system. The carbonyl (C=O) groups are reduced under alkaline conditions to enolate ($-O^-Na^+$) groups which give the leuco form water solubility.

Vat dyes are thus exclusively of the carbonyl chemical class. Dyes of

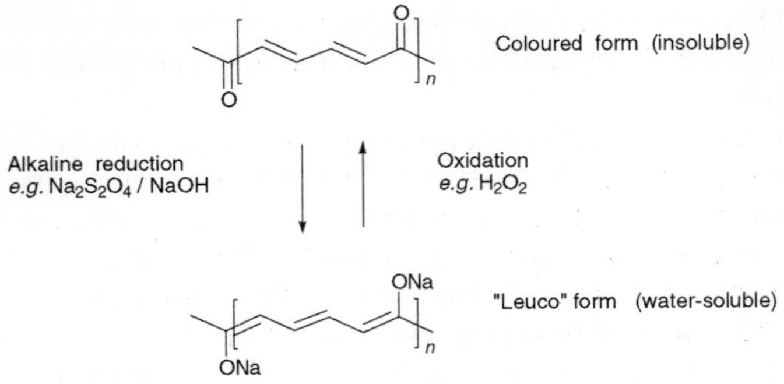

Figure 7.5   *The chemistry of vat dyeing*

other chemical classes, including azo dyes, are generally inappropriate as vat dyes because they undergo a reduction that cannot be reversed. Vat dyes are generally large planar molecules, often containing multiple ring systems, to provide the leuco form of the dyes with some affinity for the cellulosic fibre as a result of van der Waals' and dipolar forces. There is usually a notable absence of other functional groups in the molecules, because such groups can be sensitive to the reduction and oxidation reactions, although halogen substituents (Cl, Br) are encountered on occasions. Pyranthrone (**154a**, C. I. Vat Orange 9) and flavanthrone, (**154b**, C. I. Vat Yellow 1) provide examples respectively of carbocyclic and heterocyclic anthraquinone vat dyes (these carbonyl dyes are also discussed in Chapter 4). Indigo (**155**) is a further example of a long-established vat dye used commonly in the dyeing of denim fabric.

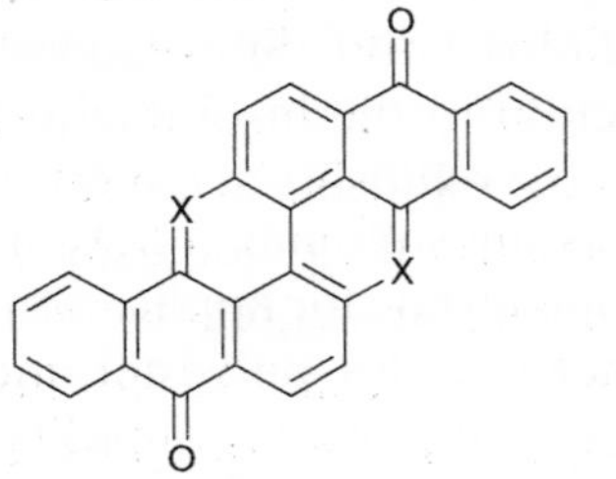

154a X = CH
154b X = N

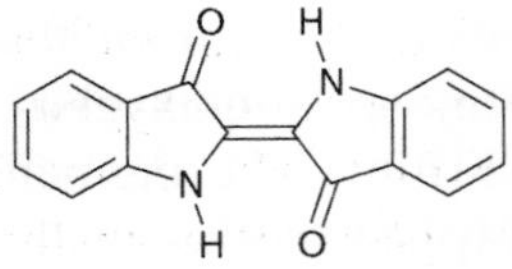

155

Azoic dyeing of cellulosic fibres is a process that is used only to a small extent today. In this process, an azo pigment is formed by chemical reaction within the fibre. The cotton fibres are first impregnated with an appropriate coupling component such as the anilide of 3-hydroxy-2-naphthoic acid, **156**, under aqueous alkaline conditions. The fibre is then treated with a solution of a stabilised diazonium salt, in which the counter-anion is tetrafluoroborate (BF_4^-) or tetrachlorozincate ($ZnCl_4^{2-}$), to form the insoluble azo pigment aggregates, which are trapped mechanically within the cotton fibres.

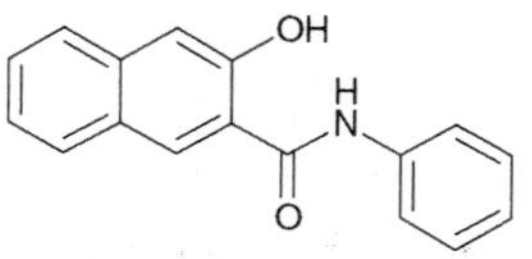

156

DYES FOR SYNTHETIC FIBRES

The three most important types of synthetic fibres used commonly as textiles are polyester, polyamides (nylon) and acrylic fibres. Polyester and the semi-synthetic fibre cellulose acetate are dyed almost exclusively with the use of disperse dyes. Polyamide fibres may be coloured using either acid dyes, the principles of which have been discussed in the section on protein fibres, or with disperse dyes. Acrylic fibres are dyed mainly using basic (cationic) dyes.

Polyester (polyethylene terephthalate, PET) has the chemical structure shown in Figure 7.6. Polyester is relatively hydrophobic (non-polar) in character, certainly in comparison to the natural protein and cellulosic fibres, largely as a result of the prominence of the benzene rings and the $-CH_2CH_2-$ groups. However, the ester groups do confer a degree of polarity on the molecule, so that the fibres are not as hydrophobic as, for example, hydrocarbon polymers such as polyethylene and polypropylene. Nevertheless, it is not surprising that relatively hydrophobic fibres such as polyester and cellulose acetate show little affinity for dyes that are ionic in character. This means that dye application classes containing ionic water-solubilising groups such as the sulfonate ($-SO_3^-$) group, for example acid and direct dyes, are inappropriate for application to polyester. The inevitable consequence is that dyes for polyester and cellulose acetate cannot be expected to have substantial water solubility. A further feature of polyester is that it is a highly crystalline fibre that consists of tightly packed, highly ordered polymer molecules. As a result, it is relatively inaccessible even to small molecules and it is thus very difficult to dye. Disperse dyes are an application class of dyes of relatively low water solubility, which may be applied as a fine dispersion in water to these relatively hydrophobic synthetic fibres. They were originally developed for application to cellulose acetate, but they have assumed much greater importance for application to polyester, and to a lesser extent to polyamides.

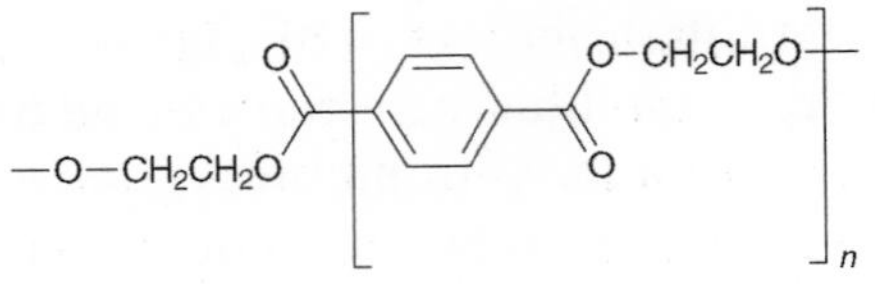

Figure 7.6   *Structure of polyester*

Disperse dyes are required to be relatively small, planar molecules to allow the dyes to penetrate between the polymer chains and into the bulk of the fibre. These dyes are commonly applied to the fibre as a fine aqueous dispersion at temperatures of around 130 °C under pressure. At these temperatures, the tight physical structure of the polymer is loosened by thermal agitation, which reduces the intermolecular bonding and facilitates entry of the dye molecules. Disperse dyes are non-ionic molecules which effectively dissolve in the polyester. In the solid solutions formed, the dye–fibre affinity is generally considered to involve a combination of van der Waals' and dipolar forces and hydrogen bonding. A general feature of disperse dye molecules is that they possess a number of polar, though not ionic, groups. Commonly encountered polar groups include the nitro (NO_2), cyano (CN), hydroxy, amino, ester, amide (NHCO) and sulfone (SO_2) groups. These polar groups can be thought of as making a number of contributions to the application properties of disperse dyes. One of their roles is to provide an adequate degree of water solubility at the high temperatures at which the dyes are applied. A second function of the polar groups is to enhance affinity as a result of dipolar intermolecular forces with the ester groups of the polyester molecule. Disperse dyes can therefore be considered as compromise molecules that possess a balanced degree of polar (hydrophobic) and non-polar (hydrophilic) character, similar to that of the polyester molecule. The polar groups will also influence the colour of the dye molecules as discussed in Chapter 2.

A consequence of the fact that disperse dyes are relatively small, non-polar molecules is that they may have a tendency to be volatile, and hence prone to sublimation out of the fibre at high temperatures. This can lead to a loss of colour and the possibility of staining of adjacent fabrics when dyed polyester is subjected to high temperatures, *e.g.* in heat setting and ironing. Increasing the size and/or the polarity of the dye molecules enhances fastness to sublimation. However, as a consequence, the compromise arises that these larger, more polar dye molecules will require more forcing conditions, such as higher temperatures and pressures, to enable the dyes to penetrate into the fibre.

The chemical structures of some typical disperse dyes are illustrated in

Figure 7.7. Numerically, azo dyes form by far the most important chemical class of disperse dyes. Azo disperse dyes may be classified into four
broad groupings. The most numerous are the aminoazobenzenes which
provide important orange, red, violet and blue disperse dyes. They are
exemplified by C. I. Disperse Orange 25 (**157**), C. I. Disperse Red 90 (**158**)
and C. I. Disperse Blue 165 (**159**). A comparison of these three
aminoazobenzene dyes provides an illustration of the bathochromic shift

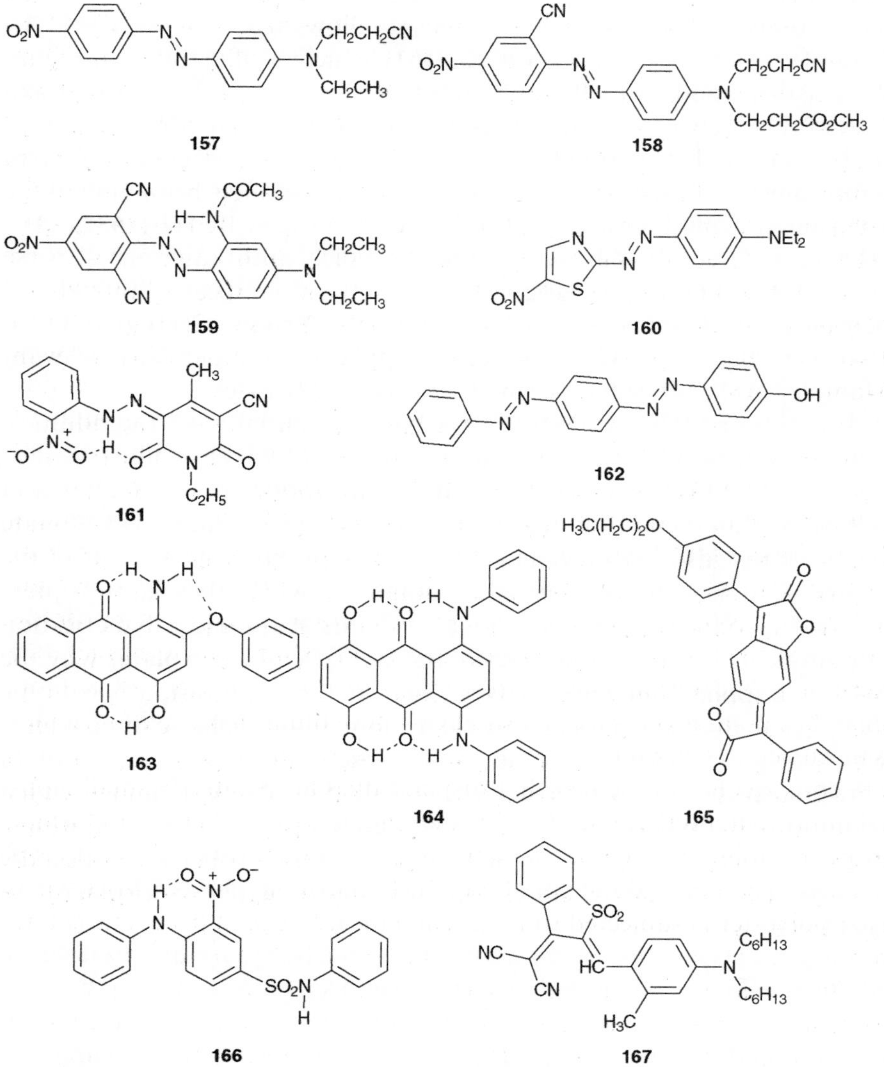

Figure 7.7 *Some examples of structures of disperse dyes*

provided by increasing the number of electron-accepting and electron-donating groups in appropriate parts of the molecules. There are two further groups of disperse dyes that are heterocyclic analogues of the aminoazobenzenes. Derivatives based on heterocyclic diazo components provide bright intense colours and are bathochromically shifted so that they serve the purpose of extending the range of blue azo disperse dyes available. An example of such a product is C. I. Disperse Blue 339 (**160**). Derivatives based on heterocyclic coupling components are useful for their ability to provide bright intense yellow azo disperse dyes. An example is C. I. Disperse Yellow 23, (**161**), which, as illustrated in Figure 7.7, exists as the ketohydrazone tautomer. The fourth group are disazo dyes of relatively simple structures, for example C. I. Disperse Yellow 23 (**162**). Carbonyl disperse dyes, especially anthraquinones, are next in importance to the azo dyes and there are a few products belonging to the nitro and polymethine chemical classes. C. I. Disperse Red 60 (**163**) and C. I. Disperse Green 5 (**164**) are examples of typical anthraquinone disperse dyes, while compound **165** is an example of the more recently-introduced benzodifuranone carbonyl type. C. I. Disperse Yellow 42, (**166**) and C. I. Disperse Blue 354, (**167**) respectively provide commercially relevant examples of the nitro and polymethine chemical classes.

Acrylic fibres are synthetic fibres based essentially on the addition polymer polyacrylonitrile, the basic structure of which is illustrated in Figure 7.8. However, most acrylic fibres are rather more complex and contain within their structure anionic groups, most commonly sulfonate ($-SO_3^-$), but also carboxylate ($-CO_2^-$) groups either as a result of the incorporation of co-polymerised monomers in which these groups present, or due to the presence of residual amounts of anionic polymerisation inhibitors. The anionic character of these acrylic fibres explains why the principal application class of dyes used for their coloration is cationic dyes. These dyes are classified by the Colour Index as basic dyes, a term which originated from their use, now largely obsolete, to dye protein fibres, such as wool, from a basic or alkaline dyebath under which conditions the protein molecules acquired a negative charge. Cationic dyes are found to exhibit rather poor fastness properties, especially lightfastness, on natural fibres but give much better performance on acrylic fibres.

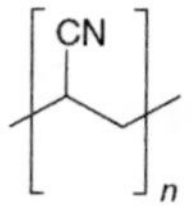

Figure 7.8   *Structure of polyacrylonitrile*

Some examples of the structures of cationic dyes used to dye acrylic fibres are shown in Figure 7.9. These include the azo dye, C. I. Basic Red 18 (**168**), the arylcarbonium ion (triphenylmethine) dye, C. I. Basic Green 4 (**169**) and the methine derivative, C. I. Basic Yellow 11, (**170**). As the name implies, the dyes are coloured cationic species, generally as a result of the presence of positively-charged quaternary nitrogen atoms (as $-NR_3^+$, or $=NR_2^+$). In the case of dyes **168** and **170** the positive charge is localised on the nitrogen atom, whereas in dye **169** it is delocalised by resonance. These groups serve two purposes. Firstly, they provide the water solubility necessary for the application of the dyes, due to their ionic character. Secondly, they provide affinity for the acrylic fibres as a result of ionic attraction between the dye cations and the anionic groups ($-SO_3^-$ and $-CO_2^-$) which are present in the acrylic fibre polymer molecules. In a sense, the means of attachment of cationic dyes to acrylic fibres may be considered as the converse of that involved in the acid dyeing of protein fibres, discussed previously in this chapter, which involves the attraction of dye anions to cationic sites on the fibre.

The two most important polyamide fibres are nylon 6.6 (**171**) and nylon 6 (**172**) whose structures are illustrated in Figure 7.10. A comparison with Figure 7.1 reveals the structural analogy between natural protein fibres such as wool and polyamide fibres. Polyamides may be dyed using acid

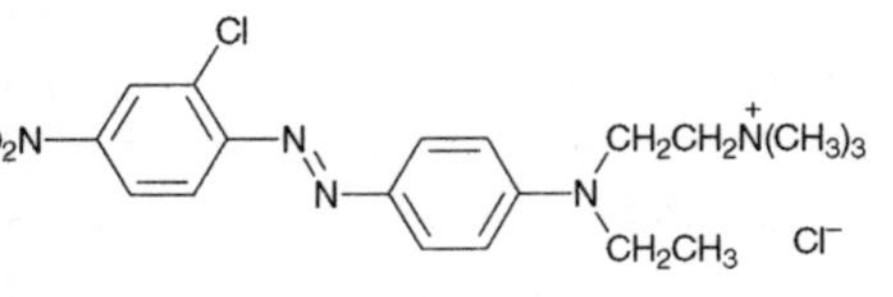

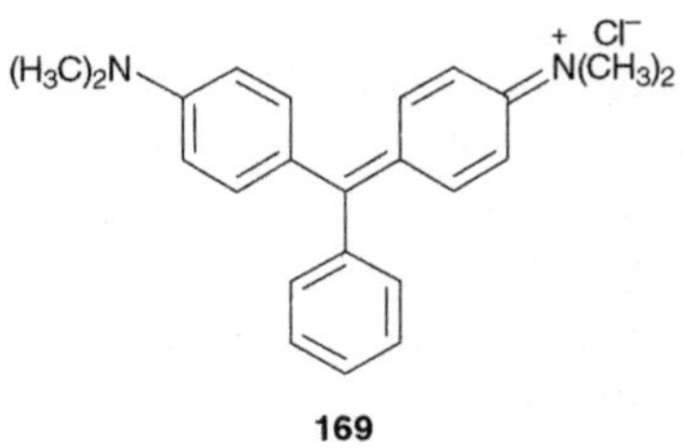

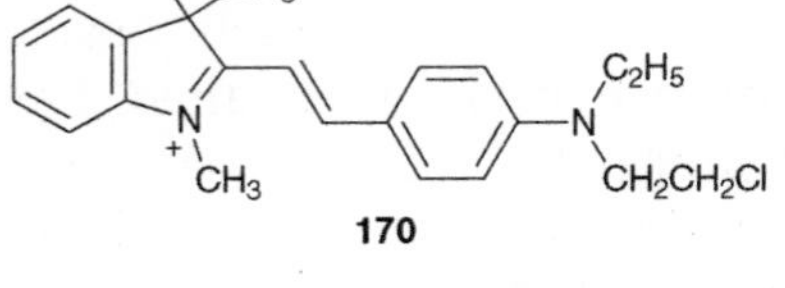

Figure 7.9 *Some examples of structures of cationic dyes for acrylic fibres*

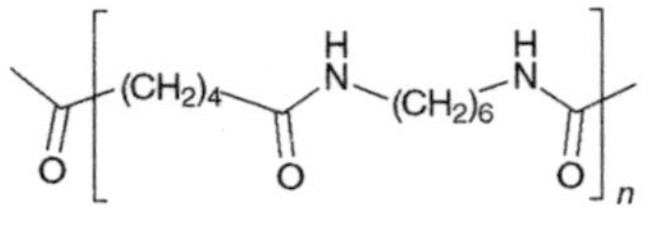

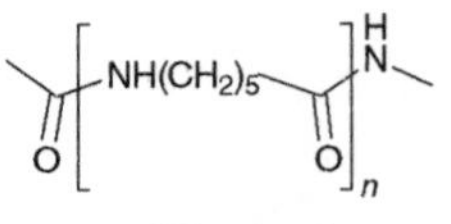

Figure 7.10 *Structure of polyamide fibres*

dyes. These are attracted to the fibres by a mechanism similar to the acid dyeing of wool, involving attraction of the dye anions to amino groups, for example at the end of the polyamide chains, which are protonated under acidic conditions. Alternatively, because polyamide fibres are relatively hydrophobic, they may be dyed using disperse dyes by a mechanism analogous to the dyeing of polyester.

Chapter 8

Reactive Dyes for Textile Fibres

It is probable that history will judge the development of reactive dyes to have been the most significant innovation in textile dyeing technology of the 20th century. As a consequence of their particular importance, and because they make use of some interesting organic chemistry, this chapter is devoted entirely to a consideration of the chemical principles involved in the application of reactive dyes. Reactive dyes, after application to the fibre, are induced to react chemically to form a covalent bond between the dye and the fibre. This covalent bond is formed between a carbon atom of the dye molecule and an oxygen, nitrogen or sulfur atom of a hydroxy, amino or thiol group on the polymer. Because of the strength of the covalent bond, reactive dyes once applied to the textile material resist removal and as a consequence show outstanding washfastness properties. Initially, reactive dyes were introduced commercially for application to cellulosic fibres, and this remains by far their most important use, although dyes of specific types have also been developed for application to protein and polyamide fibres. The potential to apply reactive dyes to other fibre types, including polyester and polypropylene, has been demonstrated technically although they are not as yet a commercial reality.

The concept of attempting to link dye molecules covalently to fibre molecules to produce colours with superior fastness to washing had been envisaged long before the 1950s when the major breakthrough took place. Prior to this, however, attempts at dye fixation employed rather vigorous conditions under which serious fibre degradation occurred. Success was first achieved in 1954, when Rattee and Stephen demonstrated that dyes containing the 1,3,5-triazinyl group were capable of reaction with cellulosic fibres under mildly alkaline conditions and under these conditions no significant degradation of the fibre was observed. The launch by ICI of Procion dyes based on this reactive system proved to be an almost immediate success. The new dyes were capable of providing superior washfastness compared with direct dyes, a much wider range of brilliant colours than were available from vat and azoic dyes, and they were capable of both continuous and batchwise application. This success provided the impetus for a considerable body of research by most dye manufacturers into the development of alternative types of fibre-reactive groups.

A general schematic representation of the structure of a reactive dye is illustrated in Figure 8.1. In the figure, four important structural features of the molecules may be identified separately: the chromogen, the water-solubilising group, the bridging group and the fibre-reactive group. The chromogen is that part of the molecule that essentially gives the molecule its colour and may contribute to other features of the dye such as its lightfastness. As encountered in most of the other application classes of textile dyes, these chromogens typically belong to the azo, carbonyl or phthalocyanine chemical classes. Specific examples are used to illustrate this feature later in this chapter. Reactive dyes are required to be water-soluble for their application to the textile fibres, and they invariably contain for this purpose one or more ionic groups, most commonly the sulfonate group as the sodium salt. They are thus anionic in nature and, as such, show many of the chemical features of acid dyes for protein fibres (Chapter 7). Most commonly, the water-solubilising group is located in the chromogenic part of the reactive dye molecule, although on occasions it is part of the fibre-reactive group. The essential structural characteristic of a reactive dye is a functional group that is capable of reacting chemically with the fibre. This feature is, for obvious reasons, termed the fibre-reactive group and the organic chemistry underlying the reaction of these groups with functionality on the fibre forms a substantial part of this chapter. Commonly the term *bridging group* is used to identify the group of atoms which is used to link the chromogenic part of the molecule to the fibre-reactive group. In many dyes, the bridging group is the amino (–NH–) group, often for reasons of synthetic convenience. There are,

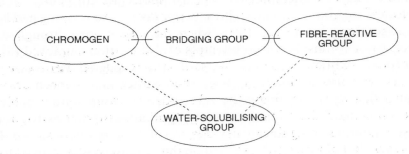

Figure 8.1   *General structure of a reactive dye*

however, certain types of reactive dyes in which no obvious bridging functionality may be clearly identified.

FIBRE-REACTIVE GROUPS

Cotton fibres are based on cellulose, a polysaccharide, whose structure is given in Figure 7.4. Most reactive dyes for cotton utilise the ability of the many hydroxy (OH) groups present in the cellulose molecule to act as nucleophiles. The dyes are commonly induced to react with the cellulose in aqueous alkali under which conditions deprotonation of the hydroxy groups (Cell–OH) takes place. This produces the more powerfully nucleophilic cellulosate anions (Cell–O⁻) which are generally regarded as the active nucleophiles in the reactive dyeing of cellulose. In the reactive dyeing of protein fibres, such as wool or silk, the nucleophilic group on the fibre may be the amino (–NH$_2$), hydroxy (–OH) or thiol (–SH) groups present in the amino side-chains of the polypeptide (see Figure 7.1). The most common types of fibre-reactive dyes for cellulosic fibres react either by aromatic nucleophilic substitution or by nucleophilic addition to alkenes.

Fibre-reactive Groups Reacting by Nucleophilic Substitution

A characteristic feature of the chemistry of aromatic ring systems is their tendency to undergo substitution reactions in which the aromatic character of the ring is retained. Since most aromatic systems are electron-rich in nature, by virtue of the system of π-electrons, the most frequently encountered reactions of aromatic compounds are electrophilic substitution reactions. Nucleophilic aromatic substitution reactions are less commonly encountered. For reaction with nucleophiles under mild conditions, aromatic systems require the presence of features which reduce the electron density of the π-system to facilitate nucleophilic attack. For example, chlorobenzene undergoes nucleophilic substitution (*e.g.* of Cl by OH) only under vigorous conditions. In contrast, 1-chloro-2,4-dinitrobenzene undergoes nucleophilic substitution very readily because of the activating effect of the two powerfully electron-withdrawing nitro groups. Reactions of this type are used in the synthesis of some nitrodiphenylamine dyes (see Scheme 6.5 in Chapter 6). Another feature of nucleophilic substitution reactions illustrated by this particular reaction is the requirement for a good leaving group, such as the chloride anion.

By far the most important fibre-reactive groups which react by nucleophilic substitution contain six-membered aromatic nitrogen-containing heterocyclic rings with halogen substituents. The first group of commercial reactive dyes, the Procion dyes, which remain among the most important in use today, contain the 1,3-5-triazine (or *s*-triazine) ring, a six-membered ring of alternate carbon and nitrogen atoms, containing at least one chlorine substituent, and with the amino (–NH–) group as the bridging group. The reaction of a chlorotriazinyl dye **173** with the cellulosate anion, incorporating an outline mechanism for the reaction which is characteristic of aromatic nucleophilic substitution, is illustrated in Scheme 8.1. A number of features may be identified as responsible for facilitating the nucleophilic substitution reaction. Firstly, the electron-withdrawing nature of the heterocyclic nitrogen atoms and, to a lesser extent, of the chlorine atom reduces the electron density in the aromatic ring which is thereby activated towards nucleophilic attack. An important feature is the extensive resonance stabilisation of the anionic intermediate **174**. In particular, it is of special importance that the delocalised negative charge is favourably accommodated on the electronegative heterocyclic nitrogen atoms in each of the three contributing resonance forms. In the final step of the sequence, the chloride ion, a particularly good leaving group, is eliminated. As a result of the reaction, a dye–fibre covalent (C–O) bond is formed, as illustrated for structure **175a**, and it is to this that the excellent fastness to wet-treatments may be attributed.

Both dichloro- and monochlorotriazinyl dyes are used commercially. Dichlorotriazinyl dyes (marketed as Procion M dyes), **173a**, are more reactive than the monochlorotriazinyl (Procion H) types **173b**. This may be explained by the fact that, in the monochloro dyes **173b**, the group X is electron-releasing, commonly amino (primary or secondary) or methoxy, replacing the electron-withdrawing chloro group in the dichloro derivatives (**173a**). There is therefore an increase in the electron density of the triazine ring in the monochlorotriazinyl dyes, which reduces their reactivity towards nucleophilic attack. Procion M dyes are capable therefore of reaction with cellulosic fibres at lower temperatures than is required for the Procion H dyes. Procion M dyes are thus applied at temperatures of around 40–60 °C whereas Procion H dyes are applied at around 80–90 °C.

A feature of the chemistry of triazinyl reactive dyes, which is in fact common to all reactive dye systems, is that they undergo, to a certain extent, a hydrolysis reaction that involves reaction of the dye with OH⁻ anions present in the aqueous alkaline dyebath in competition with the dye–fibre reaction. The hydrolysis reaction is also illustrated in Scheme 8.1. Reactive dye hydrolysis is a highly undesirable feature of reactive dyeing for a variety of reasons. In the first instance, the hydrolysed dye **175b** which is formed is no longer capable of reacting with the fibre and so must be washed out of the fibre after dyeing is complete, to ensure the

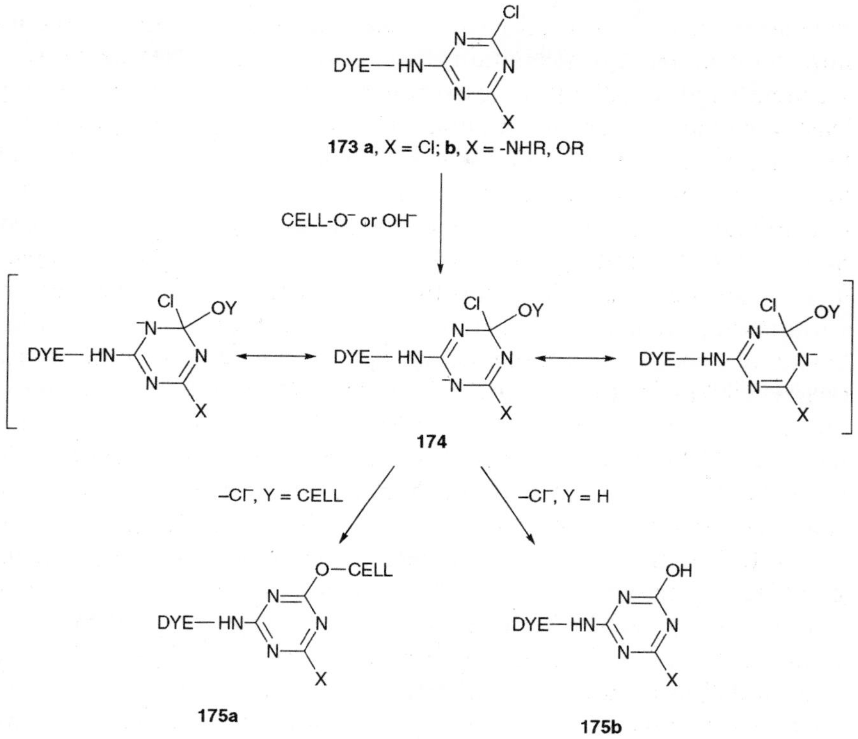

Scheme 8.1 *Reaction of Procion dyes with cellulose*

excellent fastness to washing. A second reason is that a degree of fixation less than 100% and the need for the wash-off after treatment reduces considerably the cost-effectiveness of the process. Finally, the hydrolysed dye and any unreacted dye inevitably ends up in the dyehouse effluent, and this presents environmental concerns to which society is becoming increasingly sensitive (Chapter 11).

The initial success of the reactive dyes based on the triazine ring system was immediately followed by intense research activity into the possibilities offered by other related nitrogen-containing heterocyclic systems. Numerous systems have been patented as fibre-reactive groups although only a few of these have enjoyed significant commercial success. Some examples are illustrated in Figure 8.2. They include the trichloropyrimidines **176**, the dichloropyridazines **177**, the dichloroquinoxalines **178** and the chlorobenzothiazolyl dyes **179**.

Perhaps the best known of these related groups of reactive dyes are those which utilise the 2,4,5-trichloropyrimidinyl (2,4,5-trichloro-1,3-diazinyl) system, **176**, the basis of the Drimarene dyes marketed originally in

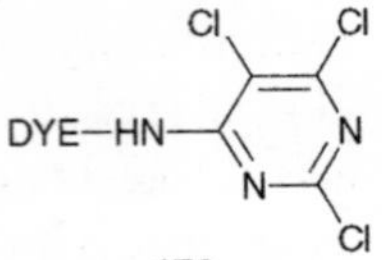

Figure 8.2 *Structures of some other heterocyclic reactive dye systems*

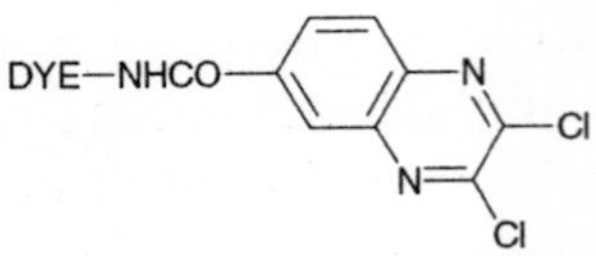

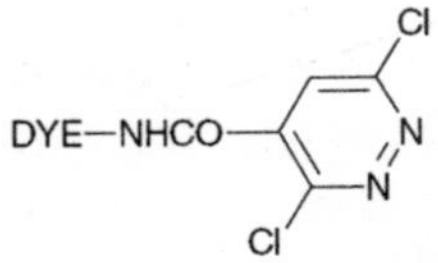

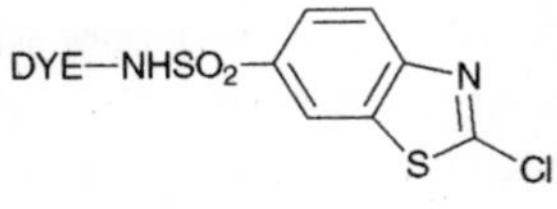

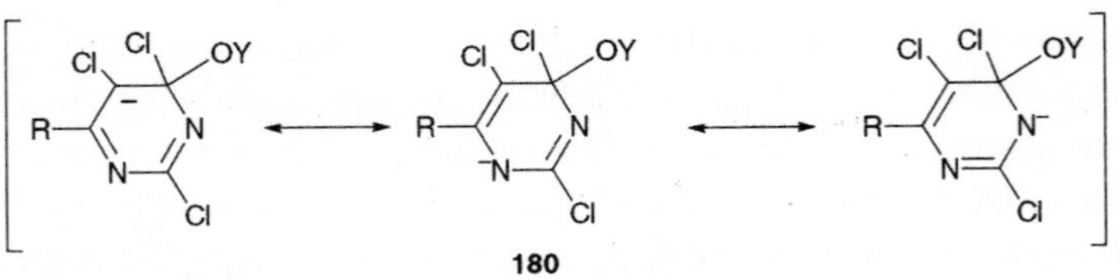

Figure 8.3 *Structure of the intermediates formed from the reaction of trichloropyrimidinyl dyes with nucleophiles*

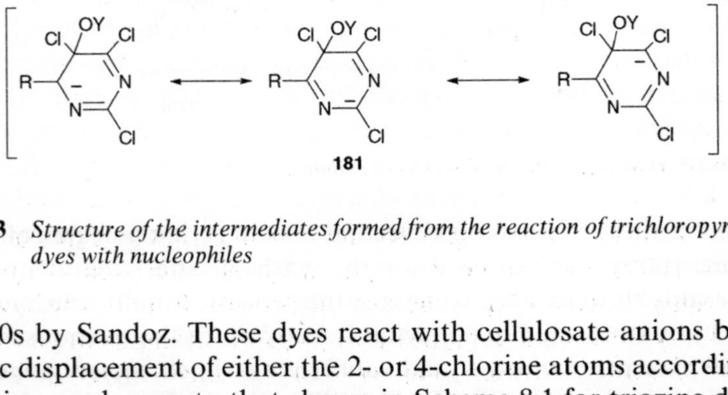

the 1960s by Sandoz. These dyes react with cellulosate anions by nucleophilic displacement of either the 2- or 4-chlorine atoms according to a mechanism analogous to that shown in Scheme 8.1 for triazine dyes. In contrast, the chlorine atom at the 5-position is not readily substituted. The reasons for this are immediately apparent from examination of the resonance forms of the intermediates derived formally from attack of the nucleophile at the relevant carbon atom as illustrated in Figure 8.3. There are three contributing resonance forms in each case. In the anionic intermediate **180**, which is formed by attack at the 4-position, two of the three contributing resonance forms have the negative charge accommodated favourably on the electronegative nitrogen atoms. The reader might like, as an exercise, to demonstrate that a similar situation arises from attack at the 2-position. In contrast, however, attack at the 5-position gives an intermediate **181** in which the negative charge is localised on carbon atoms in each of the three contributing forms. The instability of this intermediate explains the lack of reactivity at the 5-position. The 5-chloro substituent, although unreactive, is nevertheless of technological importance. Since the fibre reactive group has only two activating heterocyclic nitrogen atoms (compared with three in the Procion dyes), the 5-chloro substituent serves to enhance the reactivity of the ring towards nucleophilic attack as a result of its electron-withdrawing nature and ensures that the dyes are capable of reacting with cellulose at reasonable temperatures. It is of interest that in the cases of reactive dye systems **177–179** the bridging group is electron-withdrawing (carbonyl or sulfonyl) and no doubt plays a decisive part in activating the system to nucleophilic attack, ensuring that dyes have appropriate reactivity for the fibre.

Another method whereby the reactivity of reactive dyes may be modified involves the use of alternative leaving groups other than the chloride ion. Some examples of reactive dyes using this modification are illustrated in Figure 8.4. One of the earliest ventures in this field involved the use of quaternary amino groups, as for example in the triazine derivative **182**. Dyes using systems of this type were found to have enhanced reactivity towards cellulosic fibres, but their commercial development was inhibited by a degree of nervousness concerning the liberation of tertiary amines, pyridine in this case, into the dyebath. Some commercial success has been achieved using fluorinated heterocycles, such as those found in the triazine **183** (Cibacron F dyes, Ciba) and the pyrimidine **184** (Levafix EA dyes, Bayer) in spite of the severe hazards which needed to be overcome in their manufacture. Because of the high electronegativity of the fluorine atoms, these dyes show particularly high reactivity and are therefore capable of fixation to cellulosic fibres under mild conditions, for example at relatively low temperatures. Another leaving group which has been utilised to produce highly activated dyes is the methylsulfonyl group ($-SO_2CH_3$), used in Levafix P dyes (Bayer) which are based on the pyrimidine structure **185**.

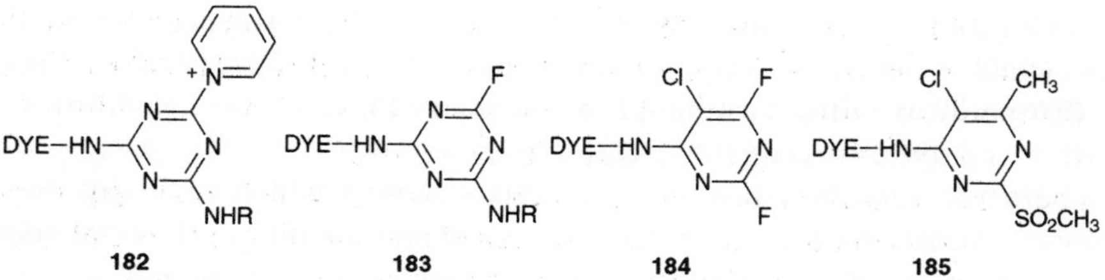

Figure 8.4 *Structures of some reactive dye systems using alternative leaving groups*

Fibre-reactive Groups Reacting by Nucleophilic Addition

Alkenes are generally regarded as relatively reactive compounds, their reactivity being attributable to the presence of the C=C double bond. Characteristically, alkenes readily undergo addition reactions, most commonly of the electrophilic type because of the electron-rich nature of the π-bond. Nucleophilic addition reactions of alkenes are less commonly encountered but can take place when there are strongly electron-withdrawing groups attached to the double bond, thereby reducing the electron density and thus facilitating nucleophilic attack. Nucleophilic addition to substituted alkenes of this type is alternatively referred to as either conjugate addition or Michael addition.

The most important reactive dyes in commercial use for application to cellulosic fibres in which the fibre-reactive groups react by nucleophilic addition are the Remazol reactive dyes. These dyes, based on the vinylsulfone reactive group, were introduced by Hoechst soon after the launch of the Procion dyes based on the triazine system by ICI. The chemistry of the process in which vinylsulfone dyes react with cellulose under alkaline conditions is illustrated in Scheme 8.2. The dye is supplied by the manufacturers as the β-sulfatoethylsulfone **186**, which is not itself fibre-reactive and is commonly referred to as the stable storage form. The water-solubilising sulfonate group, as a sulfate ester, is in the case of these dyes attached to the latent fibre-reactive group. As illustrated mechanistically in the scheme, the reaction of compound **186** with aqueous alkali causes its conversion by an elimination reaction into the highly reactive vinylsulfone form **187**. The presence of the powerfully electron-withdrawing sulfone group activates the double bond towards nucleophilic attack. Attack by the cellulosate anion on the vinylsulfone initially leads to the anionic intermediate **188** which is stabilised by resonance, with important contributions from canonical forms in which the negative charge is delocalised on to the electronegative oxygen atoms of the sulfone group. The addition reaction is completed by protonation, and the ultimate outcome is the permanent formation of a covalent dye–fibre (C–O) bond as illustrated by structure **189**. In this reactive dyeing process, as with dyes which react by nucleophilic substitution, hydrolysis also takes place in competition with the dye–fibre reaction. This reaction, leading to hydrolysed dye **190**, is also illustrated in Scheme 8.2.

There are relatively few other reactive groups which react by nucleophilic addition and which have achieved significant commercial success. However, one particular system which is worthy of note is the α-bromoacrylamido group, found in the Lanasol dyes (Ciba), which are among the most widely used reactive dyes for wool. The chemistry

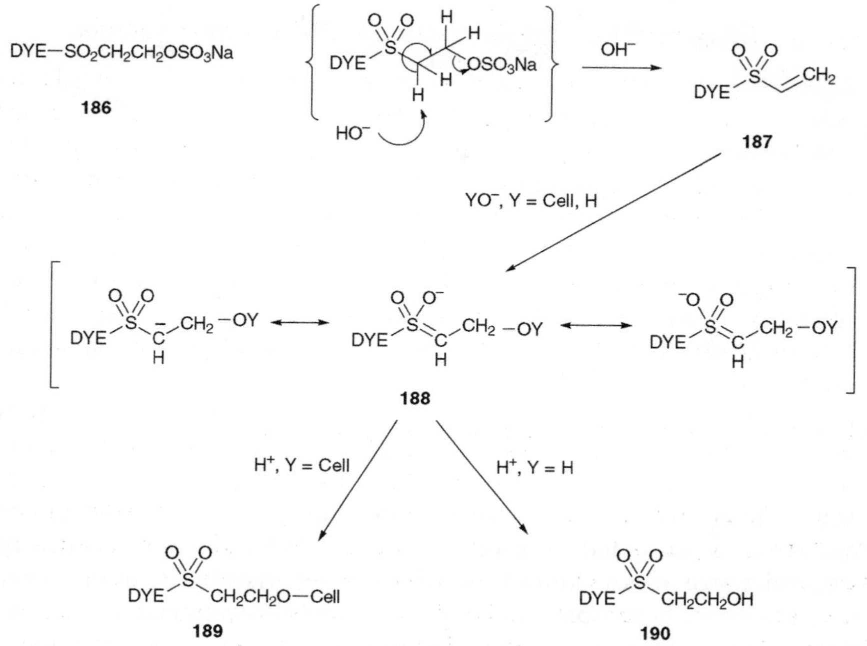

Scheme 8.2   *Reaction of vinylsulfone dyes with cellulosic fibres*

involved in the reaction of these dyes with the amino groups present on the wool fibre is outlined in Scheme 8.3. As well as the nucleophilic addition reaction, there is the possibility that the second bromine atom may be replaced by nucleophilic substitution leading, in principle, to cross-linking of the fibre, so that the dyes may be considered as bifunctional.

Bifunctional Reactive Dyes

There is no doubt that the major weakness of the reactive dyeing process is the hydrolysis reaction and the consequent need for a wash-off process. The extent to which dye hydrolysis takes place in competition with dye–fibre reaction varies quite markedly within the range 10–40% depending upon the system in question. A considerable amount of research has therefore been devoted to the search for reactive dyes with improved fixation. The most successful approach to addressing this issue has involved the development of dyes with more than one fibre-reactive group in the molecule, which statistically improves the chances of dye–fibre bond formation. Examples of products of this type are the Procion H-E

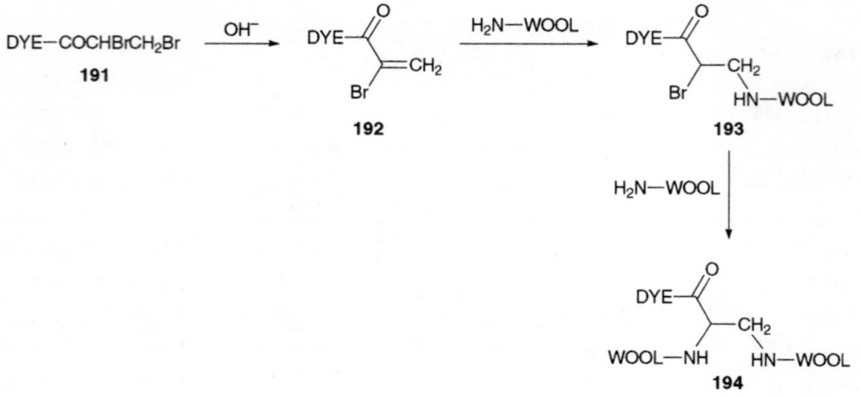

Scheme 8.3 *Reaction of α-bromoacrylamido reactive dyes with wool*

range which contain two monochlorotriazinyl reactive groups, for example as represented by general structure **195** and some dyes of the Remazol range which contain two β-sulfatoethylsulfone groups. These are examples of homo-bisfunctional dyes. Another approach utilises two different types of reactive groups, the so-called hetero-bisfunctional dyes, as for example structure **196** which contains both a monochlorotriazinyl and a β-sulfatoethylsulfone group. This particular class of reactive dye, because of the differing reactivities of the two groups, can offer greater flexibility in performance, for example reducing sensitivity to temperature and to pH variation. Although these bifunctional reactive dyes offer a significantly improved degree of fixation, the development of a practical reactive dye system which is completely free of the problems associated with hydrolysis or incomplete dye–fibre reaction remains an important target for dye chemists, and one which so far has proved elusive in spite of many years of research effort.

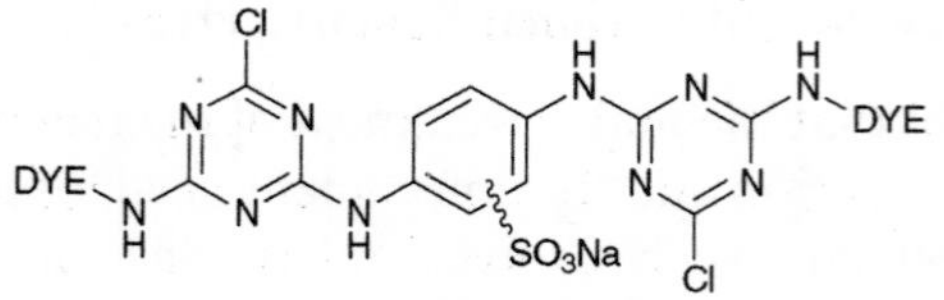

195

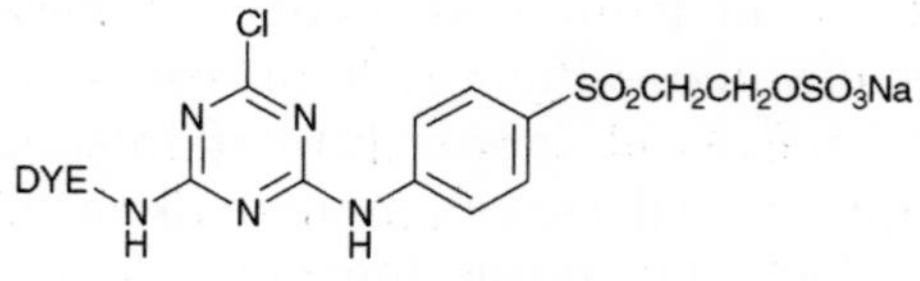

196

CHROMOGENIC GROUPS

A selection of representative chemical structures from the vast range of reactive dyes now available commercially is given in Figure 8.5. Reactive dyes may be prepared in principle from any of the chemical classes of colorant by attaching a fibre-reactive group to an appropriate molecule. In common with most application classes of textile dyes and pigments, most reactive dyes belong to the azo chemical class, especially in the yellow, orange and red shade areas. Examples are typified by the structurally related red monoazo reactive dyes, the dichlorotriazine **197a**, C. I. Reactive Red 1, the monochlorotriazine **197b**, C. I. Reactive Red 3 and the trichloropyrimidine **198**, C. I. Reactive Red 17. Bright blue reactive dyes are commonly derived from anthraquinones. Examples include the triazinyl dyes C. I. Reactive Blue 4 (**199a**), C. I. Reactive Blue 5 (**199b**) and the structurally related vinylsulfone dye, C. I. Reactive Blue 19 (**200**). Turquoise shades are produced using copper phthalocyanine derivatives, while ruby, violet and navy blue dyes commonly make use of the square planar copper complexes of appropriate azo dyes. The dioxazine system may also be used in violet and blue reactive dyes.

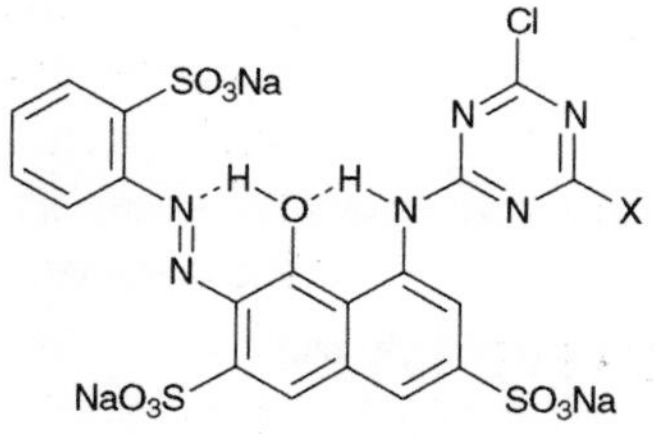

197a X = Cl; **197b** X = NHPh

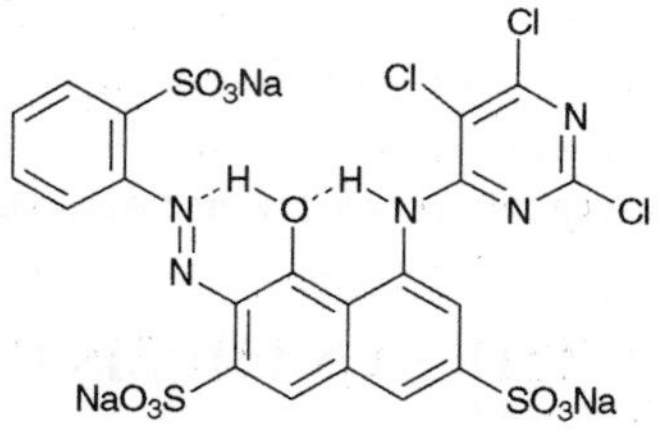

198

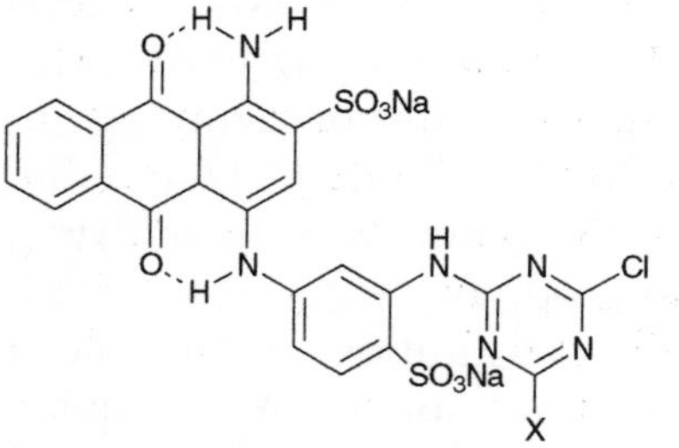

199a X = Cl;
199b X = -NH-C$_6$H$_4$-3-SO$_3$Na

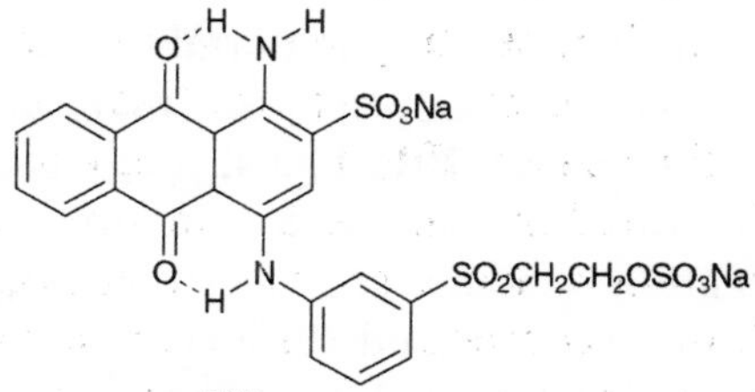

200

Figure 8.5 *Some typical chemical structures of reactive dyes*

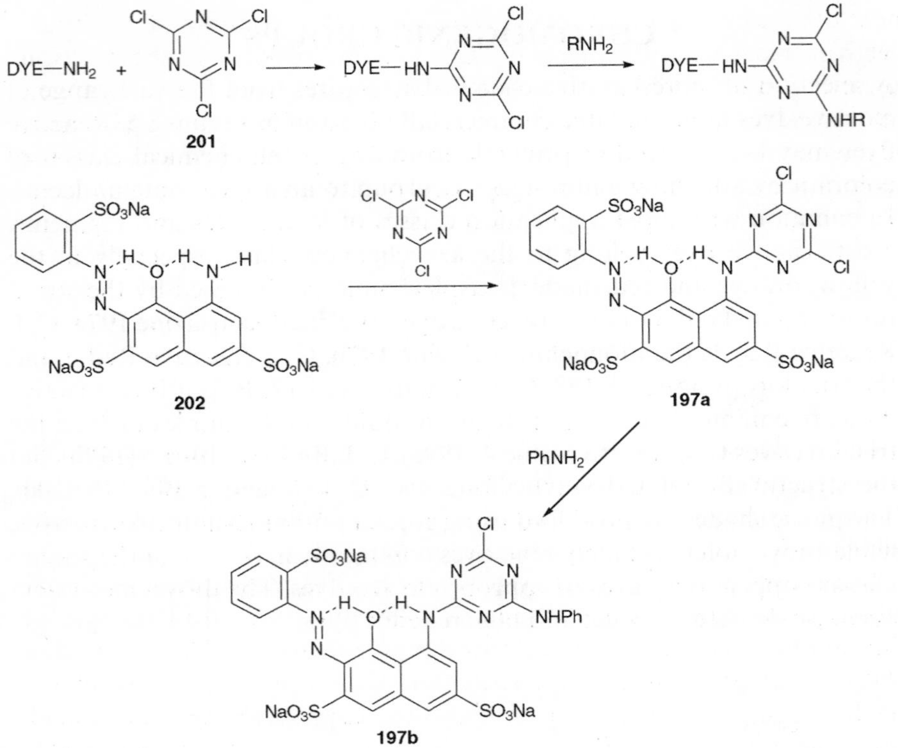

Scheme 8.4   *Synthesis of chlorotriazinyl reactive dyes*

THE SYNTHESIS OF REACTIVE DYES

The principles of the synthesis of the important chemical classes of dyes from which reactive dyes may be prepared have been discussed in Chapters 3–6 of this book. This section deals specifically with those aspects of the synthetic sequences which are used to introduce the fibre-reactive group. The starting material for the synthesis of chlorotriazinyl reactive dyes is the highly reactive material cyanuric chloride (2,4,6-trichloro-1,3,5-triazine), **201**. The strategy used in the synthesis of these dyes, as illustrated in Scheme 8.4, involves at appropriate stages of the overall reaction scheme, sequential nucleophilic substitution of the chlorine atoms of compound **201** by reaction with primary amines. As an example, the synthesis of dichlorotriazinyl dye **197a** is achieved by formation of the monoazo dye **202** from reaction of diazotised aniline-2-sulfonic acid with H-Acid under alkaline conditions, followed by its condensation with cyanuric chloride **201**. Treatment of dye **197a** with aniline under appropriate conditions gives the monochlorotriazinyl dye **197b**. Since replacement of an electron-withdrawing chlorine atom in compound **201** by an electron-releasing amino group deactivates the product of the reaction towards further nucleophilic substitution, replacement of a subsequent chlorine atom requires more vigorous conditions. This is a useful feature of the chemistry of the process since it facilitates the selectivity of the reaction sequence that leads to mono and dichlorotriazinyl reactive dyes.

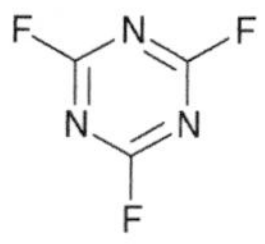

203 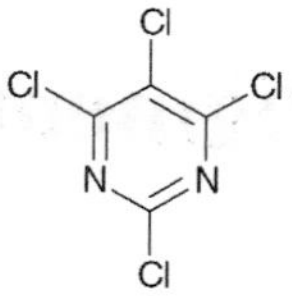   204 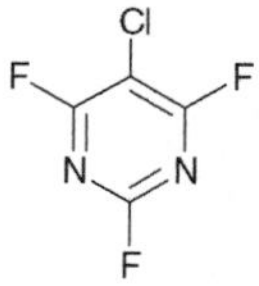   205

The syntheses of fluorotriazine, trichloropyrimidine and chlorodifluoropyrimidine dyes are completely analogous, using respectively as starting materials cyanuric fluoride (**203**), 2,4,5,6-tetrachloropyrimidine (**204**) and 5-chloro-2,4,6-trifluoropyrimidine (**205**).

The most commonly employed routes for the preparation of the β-sulfatoethylsulfone group, which is the essential structural feature of vinylsulfone reactive dyes, are illustrated in Scheme 8.5. One method of synthesis involves, initially, the reduction of an aromatic sulfonyl chloride, for example with sodium sulfite, to the corresponding sulfinic acid. Subsequent condensation with either 2-chloroethanol or ethylene oxide gives the β-hydroxyethylsulfone, which is converted into its sulfate ester by treatment with concentrated sulfuric acid at 20–30 °C. An alternative route involves treatment of an aromatic thiol with 2-chloroethanol or ethylene oxide to give the β-hydroxyethylsulfonyl compound which may then be converted by oxidation into the β-hydroxyethylsulfone.

Scheme 8.5 *Synthesis of vinylsulfone reactive dyes*

Chapter 9

Pigments

The distinction between pigments and dyes, which is based on the differences in their solubility characteristics, has been discussed in detail in Chapter 2. A pigment is a finely divided solid colouring material, which is essentially insoluble in its application medium. Pigments are used mostly in the coloration of paints, printing inks, and plastics although they are applied to a certain extent in a much wider range of substrates, including paper, textiles, rubber, glass, ceramics, cosmetics, crayons, and building materials such as cement and concrete. In most cases, the application of pigments involves their incorporation into a liquid medium, for example a wet paint or ink or a molten thermoplastic material, by a dispersion process in which clusters or agglomerates of pigment particles are broken down into primary particles and small aggregates. The pigmented medium is then allowed to solidify, either by solvent evaporation, physical solidification or by polymerisation, and the individual pigment particles become fixed mechanically in the solid polymeric matrix. In contrast to textile dyes where the individual dye molecules are strongly attracted to the individual polymer molecules of the fibres to which they are applied, pigments are considered to have only a weak affinity for their application medium, and only at the surface where the pigment particle is in contact with the medium.

Pigments are incorporated to modify the optical properties of a substrate, the most obvious effect being the provision of colour. However, this is not the only optical function of a pigment. The pigment may also be required to provide opacity, most critically in paints, which are generally designed to obscure the surface to which they are applied. Alternatively, and in complete contrast, high transparency may be important, for example in multicolour printing, which uses inks of four colours, the three subtractive primaries, yellow, magenta and cyan, together with black. In this process, transparency is essential to ensure that subsequently printed colours do not obscure the optical effect of the first colour printed. As with dyes, pigments are required to exhibit an appropriate range of fastness characteristics. They are required to be fast to light, weathering, heat and chemicals such as acids and alkalis to a degree dependent on the demands of the particular application. In addition, they are required to show solvent resistance, which refers to their ability to resist dissolving in solvents with which they may into contact in their application, to minimise problems such as 'bleeding' and migration.

In chemical terms, pigments are conveniently classified as either inorganic or organic. These two broad groups of pigments are of roughly comparable importance industrially. In general, inorganic pigments are capable of providing excellent resistance to heat, light, weathering, solvents and chemicals, and in those respects they can offer technical advantage over most organic pigments. In addition, inorganic pigments are generally of significantly lower cost than organics. On the other hand, they commonly lack the intensity and brightness of colour of typical organic pigments. Organic pigments are characterised by high colour strength and brightness although the fastness properties which they offer are somewhat variable. There is, however, a range of high-performance organic pigments which offer excellent durability while retaining their superior colour properties but these tend to be rather more expensive. The ability either to provide opacity or to ensure transparency provides a further contrast between inorganic and organic pigments. Inorganic pigments are, in general, high refractive index materials which are capable of giving high opacity while organic pigments are of low refractive index and consequently are transparent.

This chapter provides an overview of the characteristic structural features of the most important commercial pigments. If it seems that the chapter places greater emphasis on inorganic pigments, this is because the various chemical classes of organic pigments are dealt with, to a certain extent, in Chapters 3–6. In individual cases there is some discussion of structure–property relationships. Such relationships are rather more complex with pigments than with dyes, because of the dependence of the colouristic and technical performance of pigments not only on the molecular structure but also on the crystal structure arrangement and on the nature of the pigment particles, particularly their size and shape distribution. The section on inorganic pigments presents an outline of the synthetic procedures used for their manufacture. Discussion of the synthesis of organic pigments is omitted as this is dealt with in relevant earlier chapters concerned with the specific chemical classes. The manufacture of a pigment may be considered as involving two distinct phases. The first of these is the sequence of chemical reactions in which the pigment is formed. The second phase, which may run concurrently with the synthetic sequence or which may involve specific aftertreatments, ensures that the pigment is obtained in the optimum physical form. This can involve, for example, development of the appropriate crystal form, control of the particle size distribution and modification of the surfaces of the particles. Organic pigments are commonly prepared in as fine a particle size as is technically feasible to give maximum colour strength and transparency. In contrast, the particle size of many inorganic pigments is controlled carefully (often in the range 0.2–0.3 μm) to provide maximum opacity. Surface treatments are commonly used to improve the performance of pigments. For example, treatment with organic surface-active agents may lead to an improvement in the ease of dispersion into organic application media, while coating the particles with inorganic oxides, such as silica, may be used to improve the lightfastness and chemical stability of certain inorganic pigments.

INORGANIC PIGMENTS

Natural inorganic pigments, derived mainly from mineral sources, have been used as colorants since prehistoric times and a few, notably iron oxides, remain of some significance today. The origin of the synthetic inorganic pigment industry may be traced to the rudimentary products produced by the ancient Egyptians, pre-dating the synthetic organic colorant industry by several centuries (Chapter 1). The range of modern inorganic pigments was developed for the most part during the 20th century and encompasses white pigments, by far the most important of which is titanium dioxide, black pigments, notably carbon black, and coloured pigments of a variety of chemical types, including oxides (*e.g.* of iron and chromium), cadmium sulfides, lead chromates and the structurally more complex ultramarine and Prussian blue.

The structural chemistry and properties of the important chemical types of inorganic pigments are dealt with in the sections which follow, together with an outline of the most important synthetic methods. The colour of inorganic pigments arises from electronic transitions which are quite diverse in nature and different from those responsible for the colour of organic colorants. For example, they may involve charge transfer transitions, either ligand–metal (*e.g.* in lead chromates) or between two metals in different oxidation states (in Prussian blue). In ultramarines the colour is due to radical anions trapped in the crystal lattice. Inorganic pigments generally exhibit high inherent opacity, a property which may be attributed to the high refractive index which results from the compact atomic arrangement in their crystal structure. Various synthetic methods are employed in the manufacture of inorganic pigments. Frequently, the chemistry is carried out in aqueous solution from which the pigments can precipitate directly in a suitable physical form. In some cases, high temperature solid state reactions are used (*e.g.* mixed phase oxides, ultramarines), while gas-phase processes because of their suitability for continuous large-scale manufacture are of importance for the manufacture of the two largest tonnage pigments, *viz.* titanium dioxide and carbon black.

Titanium Dioxide and Other White Pigments

White pigments are conveniently classified as either hiding or non-hiding types, depending on their ability to provide opacity. By far the most important white opaque pigment is titanium dioxide (TiO_2, C. I. Pigment White 6). It finds widespread use in paints, plastics, printing inks, rubber, paper, synthetic fibres, ceramics and cosmetics. It owes its dominant industrial position to its ability to provide a high degree of opacity and whiteness (maximum light scattering with minimum light absorption) and to its excellent durability and non-toxicity. The pigment is manufactured in two polymorphic forms, rutile and anatase, the former being far more important commercially. The rutile form has a higher refractive index (2.70) than the anatase form (2.55), a feature which is attributed to the particularly compact atomic arrangement in its crystal structure, and it is therefore more opaque. In addition, the rutile form is more durable.

Two processes are used in the manufacture of titanium dioxide pigments: the *sulfate* process and the *chloride* process. The chemistry of the sulfate process, the longer established of the two methods, is illustrated schematically in Scheme 9.1. In this process, crude ilmenite ore, which contains titanium dioxide together with substantial quantities of oxides of iron, is digested with concentrated sulfuric acid, giving a solution containing the sulfates of Ti(IV), Fe(III) and Fe(II). Treatment of this

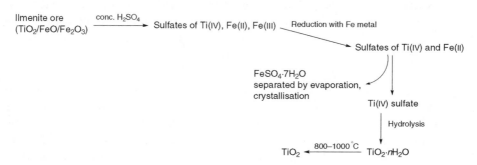

Scheme 9.1   *Sulfate process for the manufacture of TiO$_2$*

solution with iron metal then effects reduction of the Fe(III) ensuring that the iron in solution is exclusively in the Fe(II) oxidation state. The solution at this stage is then concentrated, thereby depositing crystals of $FeSO_4 \cdot 7H_2O$ (copperas), a major by-product of the process which is removed by filtration. Subsequently, in the critical step, the solution is boiled, leading to a precipitate of hydrated titanium dioxide as a result of hydrolysis of the aqueous titanium(IV) sulfate. The hydrated oxide formed is finally calcined at 800–1000 °C to remove water and residual sulfate (as H_2SO_4) and this leads to the formation of anhydrous titanium dioxide. The sulfate process may be adapted to prepare either the rutile or anatase form of the pigment, by using a 'seed' of the appropriate material at the precipitation stage.

In the more modern chloride process (Scheme 9.2), rutile titanium dioxide ore is initially treated with chlorine in the presence of carbon as a reducing agent at 800–1000 °C to form titanium tetrachloride. After purification by distillation, the tetrachloride is subjected to gas-phase oxidation at 1500 °C with air or oxygen to yield a high purity, fine particle size rutile titanium dioxide pigment. Chlorine is generated at this stage and may be recycled. The two manufacturing processes are of roughly comparable importance on a worldwide basis. However, the chloride process offers certain inherent advantages over the sulfate route. These include suitability for continuous operation, excellent control of pigment properties and fewer by-products, which in the case of the sulfate process can lead to waste disposal problems.

Zinc sulfide (ZnS, C. I. Pigment White 7) and antimony(III) oxide (Sb_2O_3, C. I. Pigment White 11) are white hiding pigments which find some specialist applications but their lower refractive indices mean that they are less efficient than TiO_2 in producing opacity. White lead (basic lead carbonate, C. I. Pigment White 1), formerly the traditional white hiding pigment, has become virtually obsolete on the grounds of both inferior technical performance and toxicity.

Non-hiding white pigments, sometimes referred to as extenders or

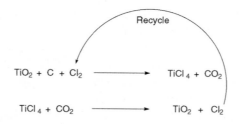

Scheme 9.2   *Chloride process for the manufacture of TiO_2*

fillers, are low cost products used in large quantities particularly by the plastics industry. They are white powders of relatively low refractive index and thus they are capable of playing only a minor role in providing opacity. They are, however, used in a variety of other ways. For example, they may be used to modify the flow properties of paints and inks and to modify the mechanical properties and lower the cost of plastics. Commonly used non-hiding white pigments include calcium carbonate ($CaCO_3$), barium sulfate ($BaSO_4$), talc (hydrated magnesium silicate), china clay (hydrated aluminium silicate) and silica.

Coloured Oxides and Oxide-hydroxides

By far the most important coloured inorganic pigments are the iron oxides, which provide colours range from yellow and red to brown and black. They are used extensively in paints, plastics and in building materials such as cement and concrete. Both natural and synthetic iron oxide pigments are used commercially. Oxides of iron are major constituents of some of the most abundant minerals in the earth's crust. Natural iron oxide pigments are manufactured from deposits of suitable purity by a milling process. The natural iron oxides, examples of which include yellow ochre, red hematite and burnt sienna, are cost-effective materials which meet many of the demands of the colour industry. Micaceous iron oxide is a natural pigment used in metal-protective coatings. Its flake-like particles laminate in the paint film, forming a reflective surface which reduces radiation degradation of the film, and providing a barrier to moisture as an aid to corrosion protection.

Synthetic iron oxide pigments offer the advantages over their natural counterparts of chemical purity and improved control of physical form. A number of different structural types are encountered. Red iron oxides (C. I. Pigments Red 101 and 102) consist principally of anhydrous iron(III) oxide (Fe_2O_3) in its α-crystal modification. Yellow iron oxide pigments (C. I. Pigments Yellow 42 and 43), although often formulated as hydrated iron(III) oxides, are better represented as iron(III) oxide-hydroxides, FeO(OH). The principal constituent of black iron oxide pigments is a non-stoichiometric mixed Fe(II)/Fe(III) oxide. It is usually formulated as Fe_3O_4, however, as the two oxidation states are generally present in approximately equal proportions. Brown pigments may be derived from the mixed Fe(II)/Fe(III) oxide or from mixtures containing Fe_2O_3 and FeO(OH). Iron oxide pigments are characterised, in general, by excellent durability, high opacity, low toxicity and low cost. However, the yellow pigments show somewhat lower heat stability because of their tendency to lose water at elevated temperatures, in the process turning redder due to the formation of Fe_2O_3. The colour of iron oxide pigments has been attributed principally to light absorption as a result of ligand–metal charge transfer, although probably influenced also by the presence of crystal field d–d transitions. The main deficiency of iron oxide pigments is that the colours lack brightness and intensity.

The most important synthetic routes to iron oxide pigments involve either thermal decomposition or aqueous precipitation processes. A method of major importance for the manufacture of α-Fe_2O_3, for example, involves the thermal decomposition in air of $FeSO_4 \cdot 7H_2O$ (copperas) at temperatures between 500 °C and 750 °C. The principal method of manufacture of the yellow α-$FeO(OH)$ involves the oxidative hydrolysis of Fe(II) solutions, for example in the process represented by reaction (1).

$$4FeSO_4 + 6H_2O + O_2 \rightarrow 4FeO(OH) + 4H_2SO_4 \qquad (1)$$

The reaction is sustained by addition of iron metal which reacts with the sulfuric acid formed, regenerating Fe(II) in solution. To ensure that the desired crystal form precipitates, a seed of α-$FeO(OH)$ is added. However, with appropriate choice of conditions, for example of pH and temperature and by ensuring the presence of appropriate nucleating particles, the precipitation process may be adapted to prepare either the orange–brown γ-$FeO(OH)$, the red α-Fe_2O_3 or the black Fe_3O_4.

The only other 'simple' oxide pigment of major significance is chromium(III) oxide, Cr_2O_3, C. I. Pigment Green 17. This is a tinctorially weak, dull green pigment but it shows outstanding durability, including thermal stability to 1000 °C. The pigment is normally prepared by treatment of chromates or dichromates with reducing agents such as sulfur or carbon.

The mixed phase oxides are a group of inorganic pigments which were developed originally for use in ceramics but which have subsequently found widespread application in plastics because of their outstanding heat stability and weathering characteristics combined with moderate colour strength and brightness. Structurally, the pigments may be considered to be formed from stable oxide host lattices, *e.g.* rutile (TiO_2), spinel ($MgAl_2O_4$) and inverse spinel, into which are incorporated transition metal ions, *e.g.* Cr^{3+}, Mn^{2+}, Fe^{3+}, Co^{2+}, Ni^{2+}. This provides a range of colours in which the excellent durability characteristics of the host crystal structures are retained. An important commercial example of a mixed oxide pigment based on the spinel lattice is cobalt aluminate blue (C. I. Pigment Blue 28), usually represented as $CoAl_2O_4$, although in practice it is found to contain slightly less cobalt than this formula would indicate. While the successful formation of a mixed-phase oxide requires that the 'foreign' cation must have a suitable ionic radius to be incorporated into its lattice position, a similar valency to that of the metal ion replaced is not essential. For example, a metal ion of lower valency may be incorporated into the lattice provided that an element in a higher oxidation state is incorporated at the same time in the amount required to maintain statistical electrical neutrality. As an example, nickel antimony titanium yellow (C. I. Pigment Yellow 53), an important member of the series, is derived from the rutile TiO_2 structure by partial replacement of Ti^{4+} ions with Ni^{2+} ions, at the same time incorporating antimony(v) atoms such that the Ni/Sb ratio is 1:2.

Mixed-phase oxide pigments are manufactured by high temperature (800–1000 °C) solid state reactions of the individual oxide components in the appropriate quantities. The preparation of nickel antimony titanium yellow, for example, involves reaction of TiO_2, NiO and Sb_2O_3 carried out in the presence of oxygen or other suitable oxidising agent to effect the necessary oxidation of Sb(III) to Sb(v) in the lattice.

Cadmium Sulfides, Lead Chromates and Related Pigments

Cadmium sulfides and sulfoselenides provide a range of moderately intense colours ranging from yellow through orange and red to maroon. They are of particular importance in the coloration of thermoplastics, especially in engineering polymers which are processed at high temperatures, because of their outstanding heat stability. Cadmium sulfide, CdS (C. I. Pigment Yellow 37) is dimorphic, existing in α- and β-forms. The shade range of the commercial pigments is extended by the formation of solid solutions. When cadmium ions are partially replaced in the lattice by zinc, greenish-yellow products result, whilst replacement of sulfur by selenium gives rise progressively to the orange, red and maroon sulfoselenides (C. I. Pigment Orange 20) and (C. I. Pigment Red 108), depending on the degree of replacement.

Lead chromate pigments provide a range of colours, from greenish-yellow through orange to yellowish-red. They offer good fastness properties, a remarkably high brightness of colour for inorganic pigments, and high opacity at relatively low cost. Historically, lead chromate pigments were found to exhibit a tendency to darken, either on exposure to light (due to lead chromite formation) or in areas of high industrial atmospheric pollution (due to lead sulfide formation). These early problems were overcome by the use of surface treatment with oxides, for example silica, and the modern range of pigments now offer excellent durability. The variation in shade of lead chromate pigments is achieved by the formation of solid solutions. In this respect they resemble the cadmium sulfides, although structurally the lead chromate pigments present a more complex situation by exhibiting polymorphism. The mid-shade yellow products are essentially pure $PbCrO_4$ (C. I. Pigment Yellow 34) in its most stable monoclinic crystal form. Incorporation of sulfate ions into the lattice while retaining the monoclinic crystal form gives rise to the somewhat greener lemon chromes. The greenest shades (primrose chromes) consist similarly of solid solutions of $PbCrO_4$ and $PbSO_4$ but stabilised chemically in a metastable orthorhombic crystal form. Incorporation of molybdate anions into the lattice gives rise to the orange and light red molybdate chromes (C. I. Pigment Red 104). Molybdate chromes usually also contain small amounts of sulfate ions, which are thought to play a role in promoting the formation of the appropriate crystal form.

Cadmium sulfides are prepared by aqueous precipitation processes using suitable water-soluble sources of cadmium and sulfide ions. The zinc-containing pigments are formed when appropriate quantities of soluble zinc salts are incorporated into the process, while the sulfoselenides are prepared by dissolving elemental selenium in the sulfide solution before the precipitation. Since the pigments usually precipitate from solution in the less stable β-form, an essential final step in their manufacture is a controlled calcination at 600 °C which effects the conversion into the desired α-form. Lead chromate pigments are manufactured by mixing aqueous solutions of lead nitrate and sodium chromate or sodium dichromate. The mixed phase pigments result when appropriate quantities of sodium sulfate or molybdate are incorporated into the preparation.

The use of cadmium sulfide and lead chromate pigments is limited to a considerable extent on the grounds of potential toxicity due to the presence of cadmium, lead and chromium(VI). Their use is restricted by voluntary codes of practice reinforced by legislation in certain cases, including toy finishes and other consumer paints, graphic instruments and food contact applications. In the European Community, for example, a Directive restricts the use of the cadmium sulfides in applications where they are not seen to be essential. However, at present, completely satisfactory substitutes for cadmium pigments are not available for use in certain high temperature plastics applications, especially in terms of thermal and chemical stability, while lead chromates remain by far the most cost-effective durable yellow and orange pigments. It is argued, particularly by the manufacturers of these products, that as a result of their extreme insolubility they do not present a major health hazard. Nevertheless, it seems likely that the trend towards their replacement by more acceptable inorganic and organic pigments will continue in an increasing range of applications.

Two types of inorganic pigments of relatively recent origin, bismuth vanadates and cerium sulfides, are potential replacements for the so-called 'heavy metal'-containing products. Bismuth vanadates, which can contain variable amounts of bismuth molybdate, are brilliant yellow pigments with high opacity and good durability, and are used primarily to provide bright deep yellow shades in industrial and automotive paints. The pigments are manufactured by an aqueous precipitation reaction involving bismuth nitrate, sodium vanadate and sodium molybdate. The hydrated products which result are then calcined at 600 °C to both remove water and develop the appropriate crystalline form. Cerium sulfides provide yellow, orange and red shades with excellent durability. The colours, however, are tinctorially rather weak.

Ultramarines

Of this small group of pigments, ultramarine blue (C. I. Pigment Blue 29) is the best known and by far the most important, although violet and pink pigments are also produced. Ultramarine blue offers excellent fastness to light and heat at moderate cost. Although capable of providing brilliant reddish-blue colours in application, ultramarine blue suffers from poor tinctorial strength. As an example, the pigment has less than one-tenth of the colour intensity of copper phthalocyanine, the most important organic blue pigment. A further deficiency of the pigment is rather poor resistance towards acids. Ultramarine blue pigments have a complex sodium aluminosilicate zeolitic structure. In essence, the structure consists of an open three-dimensional framework of AlO_4 and SiO_4 tetrahedra and within this framework there are numerous cavities in which are found small sulfur-containing anions together with sodium cations which maintain the overall electrical neutrality. It has been conclusively demonstrated that the radical anion S_3^- is the species responsible for the blue colour (λ_{max} 600 nm) of ultramarine blue pigments. It is both interesting and somewhat surprising that products of such high durability result when species such as S_3^-, which are otherwise quite unstable, are trapped within the ultramarine lattice.

Formerly derived from the natural mineral lapis lazuli, ultramarine blue pigments have, for more than a century, been manufactured synthetically. The materials used in the manufacture of ultramarines are china clay (a hydrated aluminosilicate), sodium carbonate, silica, sulfur and a carbonaceous reducing material such as coal tar pitch. For the manufacture of the blue pigments, the blend of ingredients is heated to a temperature of 750–800 °C over a period of 50–100 h, and the reaction mixture is then allowed to cool in an oxidising atmosphere over several days.

Prussian Blue

Prussian blue (C. I. Pigment Blue 27), known also as iron blue or Milori blue, is the longest established of all synthetic colorants still in use and retains moderate importance as a low cost blue pigment. On the basis of single crystal X-ray diffraction studies, it has been concluded that Prussian blue is best represented as the hydrated iron(III) hexacyanoferrate(II), $Fe_4[Fe(CN)_6]_3 \cdot nH_2O$. However, when precipitated in the fine particle size essential for its use as a pigment, significant and variable amounts of potassium or ammonium ions are incorporated into the product by surface adsorption or occlusion. In addition, the commercial products may contain indefinite amounts of water and they can exhibit variable stoichiometry and a degree of structural disorder. In the crystal structure of Prussian blue, the Fe(II) atoms are bonded exclusively to carbon atoms in FeC_6 octahedra and the Fe(III) atoms are bonded exclusively to the nitrogen atoms. Many of the pigmentary properties of Prussian blue have been explained on the basis of its crystal structure. For example, the extreme insolubility of the material has been attributed to the fact that the complex is polymeric as a result of the –Fe(II)–C–N–Fe(III)– bonding sequence. The colour is due to metal–metal electron transfer from an Fe(II) atom to an adjacent Fe(III) atom, a phenomenon commonly encountered in mixed oxidation state compounds of this type.

The industrial production of Prussian blue is based on the reaction in aqueous solution of sodium hexacyanoferrate(II), $Na_4Fe(CN)_6$, with iron(II) sulfate, $FeSO_4 \cdot 7H_2O$ in the presence of an ammonium salt, which results initially in the formation of the colourless insoluble iron(II) hexacyanoferrate(II) (Berlin white). Prussian blue is generated by subsequent oxidation with a dichromate or chlorate.

Carbon black

Carbon blacks (C. I. Pigment Black 6 and 7) dominate the market for black pigments, providing an outstanding range of properties at low cost and finding wide use in all the usual pigment applications. One of the most important applications for carbon black pigments is in rubber where, as well as providing the colour, they fulfil a vital role as reinforcing agents. Although carbon black is virtually always classified as an inorganic pigment, there is considerable justification for classifying the product amongst the high-performance organic pigments. For example, the nature of the bonding in carbon black is organic in nature, whilst many of its properties, especially the high absorption coefficient, are arguably more closely related to those of organic than of inorganic pigments. Carbon blacks have been described as having an imperfect graphite-like structure consisting of layers of large sheets of carbon atoms in six-membered rings which are parallel but further apart than in graphite, and arranged irregularly.

Carbon blacks are manufactured from hydrocarbon feedstocks by partial combustion or thermal decomposition in the gas phase at high temperatures. World production is today dominated by a continuous furnace black process, which involves the treatment of viscous residual oil hydrocarbons that contain a high proportion of aromatics with a restricted amount of air at temperatures of 1400–1600 °C.

ORGANIC PIGMENTS

The synthetic organic pigment industry developed towards the end of the 19th century out of the established synthetic textile dye industry. Many of the earliest organic pigments were prepared from water-soluble dyes rendered insoluble by precipitation onto colourless inorganic substrates such as alumina and barium sulfate. These products were referred to as 'lakes'. A further significant early development was the discovery and commercial introduction of a range of azo pigments, which provided the basis for the most important yellow, orange and red organic pigments currently in use. These so-called classical azo pigments offer bright intense colours although generally only moderate performance in terms of fastness properties. A critical event in the development of the organic pigment industry was the discovery, in 1928, of copper phthalocyanine. This blue pigment was the first product to offer the outstanding intensity and brightness of colour typical of organic pigments, combined with an excellent set of fastness properties, comparable with many inorganic pigments. The discovery stimulated the quest for other chemical types of organic pigment which could emulate the properties of copper phthalocyanine in the yellow, orange, red and violet shade areas. This research activity gained further impetus from the emergence of the automotive paint market and the growth of the plastics and synthetic fibres industries, applications which demanded high levels of technical performance. The range of high-performance organic pigments which has emerged includes the quinacridones, isoindolines, dioxazines, perylenes, perinones and diketopyrrolopyrroles, together with a number of improved performance azo pigments.

Organic pigments generally provide higher intensity and brightness of colour than inorganic pigments. These colours are due to the $\pi-\pi^*$ electronic transitions associated with extensively conjugated aromatic systems (Chapter 2). Organic pigments are unable to provide the degree of opacity which is typical of inorganic pigments, because of the lower refractive index associated with organic crystals. However, the combination of high colour strength and brightness with high transparency means that organic pigments are especially well suited to printing ink applications. The range of commercial organic pigments exhibit variable fastness properties which are dependent both on the molecular structure and on the nature of the intermolecular association in the solid state. Since organic molecules will commonly exhibit some tendency to dissolve in organic solvents, organic pigment molecules incorporate structural features which are designed to enhance the solvent resistance. For example, an increase in the molecular size of the pigment generally improves solvent resistance. In addition, the amide (–NHCO–) group features prominently in the chemical structures of organic pigments, because its presence enhances fastness not only to solvents, but also to light and heat, as a result of its ability to participate in strong dipolar interactions and in hydrogen bonding, both intramolecular and intermolecular. The incorporation, where appropriate, of halogen substituents and of metal ions, particularly of the alkaline earths and transition elements, can also have a beneficial effect on fastness properties. The following sections provide an overview of the more important chemical types of commercial organic pigments, together with some discussion of the structural features which determine their suitability for particular applications.

Azo pigments

Azo pigments, both numerically and in terms of tonnage produced, dominate the yellow, orange and red shade areas in the range of commercial organic pigments (Chapter 3). The chemical structures of some important classical azo pigments are shown in Figure 9.1. The structures are illustrated in the ketohydrazone form since structural studies carried out on a wide range of azo pigments have, in each case, demonstrated that the pigments in the solid state exist exclusively in this form. Many other colour chemistry texts follow the commonly used convention to illustrate them in the azo tautomeric form. Simple classical monoazo pigments such as Hansa Yellow G, **206** (C. I. Pigment Yellow 1) and Toluidine Red, **207** (C. I. Pigment Red 3) are products which show bright colours and good lightfastness, but rather poor solvent resistance. The good lightfastness of these molecules is attributed to the extensive intramolecular hydrogen-bonding in the form of six-membered rings. The inferior fast-

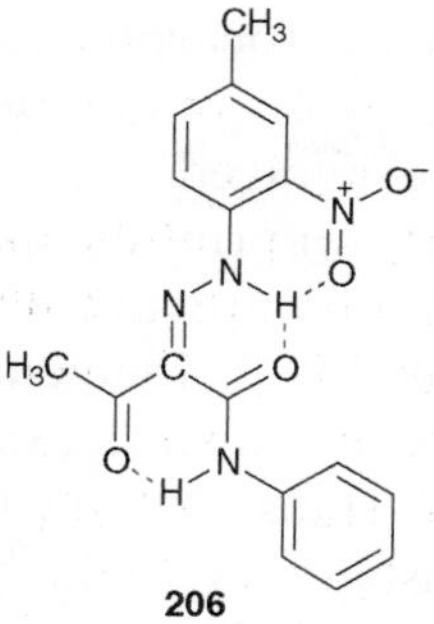

206

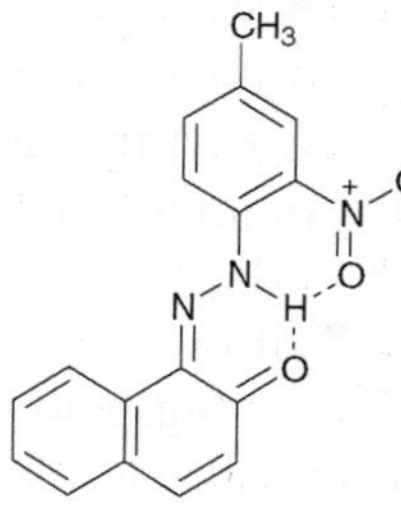

207

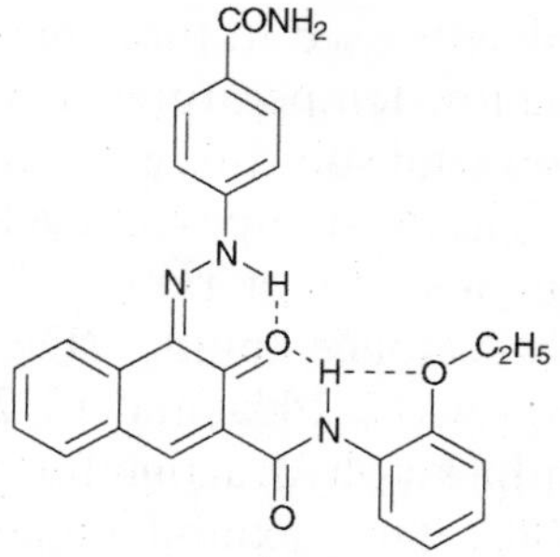

210

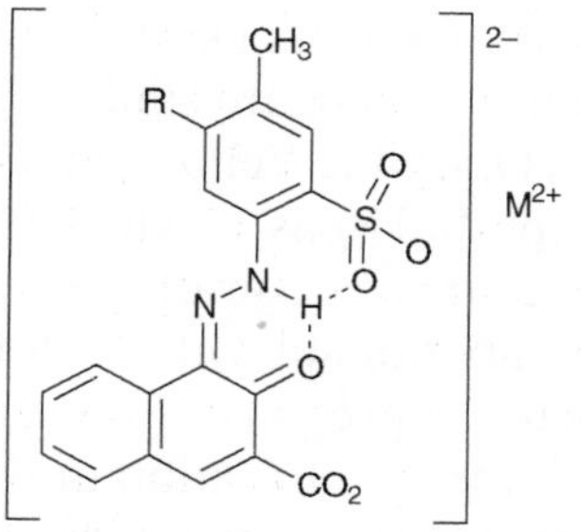

211a R = H; M = Ca
211b R = Cl; M = Ca
211c R = Cl; M = Mn

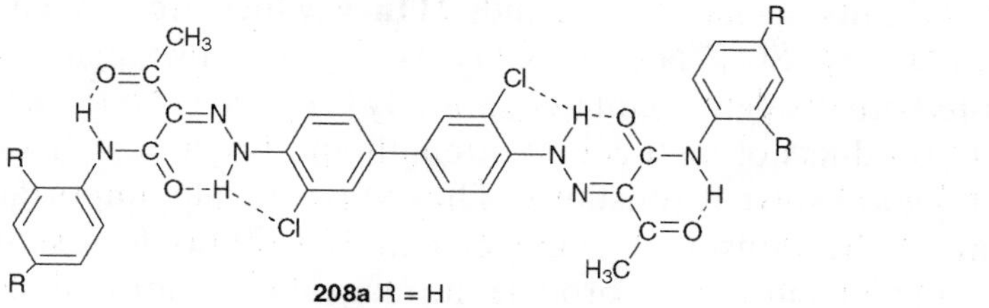

208a R = H
208b R = CH_3

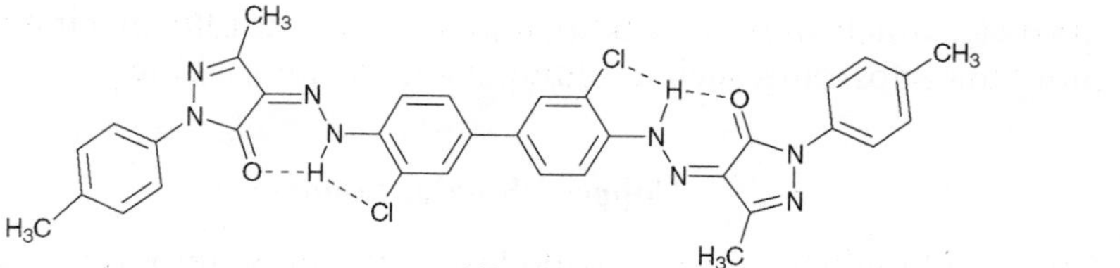

209

Figure 9.1 *Chemical structures of some important classical azo pigments*

ness to organic solvents is due to their small molecular size and the fact that the intermolecular interactions in the crystal structures involve essentially only van der Waals' forces. The use of these pigments is largely restricted to decorative paints. The pigments resist dissolving in the solvents used in these paints (either water or aliphatic hydrocarbons) at the low temperatures involved, but have a tendency to dissolve in more powerful solvents, especially if higher temperatures are involved.

The most important yellow azo pigments, particularly for printing inks but also for a range of paint and plastics applications are the disazoacetoacetanilides (Diarylide Yellows), which include C. I. Pigments Yellow 12, **208a**, and 13, **208b**. These pigments have been shown to exist in bisketohydrazone forms, structurally analogous to Hansa Yellow G (**206**). They exhibit higher colour strength and transparency than the corresponding monoazo pigments, and improved solvent fastness which is attributable to the larger molecular size. The Pyrazolone Oranges, such as C. I. Pigment Orange 34 (**209**), are similar both in chemical structure and in properties to the Diarylide Yellows and are used in a similar range of applications. The Naphthol Reds, of which C. I. Pigment Red 170 (**210**), is an important example, are structurally related to Toluidine Red (**207**). Compound **210** shows superior solvent resistance as a result of its larger molecular size and due to the presence of the amide groups which provide strong intermolecular forces of attraction in the crystal lattice structure. Consequently, it is suitable for use in a wider range of paint and plastics applications. The most important of the classical red azo pigments are metal salts, such as compounds **211a–c**, which are derived from azo dyes containing $-SO_3^-Na^+$ or $-CO_2^-Na^+$ groups by replacement of the Na^+ ions with divalent metal ions, notably Ca^{2+}, Sr^{2+}, Ba^{2+} and Mn^{2+}. They are products of high colour strength and brightness, high transparency and good solvent resistance. They are especially important for printing ink applications, C. I. Pigment Red 57:1 (**211a**), for example, being the main pigment used to provide the magenta inks for multicolour printing. Metal salt pigments have evolved from 'lake' pigments, now largely obsolete, which were essentially anionic azo dyestuffs precipitated onto inorganic substrates such as alumina and barium sulfate.

Copper Phthalocyanines

Copper phthalocyanines provide by far the most important of all blue and green pigments. The chemistry of the phthalocyanines has been discussed in some depth in Chapter 5, so only a brief account is presented here. Copper phthalocyanine, **212** (C. I. Pigment Blue 15), is arguably the single most important organic pigment. Copper phthalocyanine finds widespread use in most pigment applications because of its brilliant blue colour and its excellent resistance towards light, heat, solvents, acids and alkalis. In addition, in spite of its structural complexity, copper phthalocyanine is a relatively inexpensive pigment as it is manufactured

212

in high yield from low cost starting materials. The β-form of the pigment is almost always the pigment of choice for cyan printing inks, and it is suitable also for use in most paint and plastics applications. The most important green organic pigments are copper phthalocyanines in which most or all of the ring hydrogen atoms are replaced by halogens. These copper phthalocyanine greens, which exhibit properties comparable to the blues, are also excellent pigments.

High-performance Organic Pigments

Copper phthalocyanines, although generally regarded as classical organic pigments, exhibit outstanding technical performance and so could equally well be described as high-performance organic pigments. This section contains a brief survey of a range of the organic pigments, encompassing a wide variety of structural types, which have been developed in an attempt to match the properties of copper phthalocyanines in the yellow, orange, red and violet shades. They include two groups of azo pigments, carbonyl pigments of a variety of types and dioxazines. High-performance organic pigments are particularly suited to applications which require bright, intense colours and which at the same time place severe demands on the technical performance of pigments, such as the coatings applied to car bodies, referred to as automotive paints. They provide excellent durability, combined with good colour properties but they do tend to be rather expensive.

There are two classes of high-performance azo pigments: disazo condensation pigments and benzimidazolone azo pigments. The chemical structures of representative examples of these products are illustrated in Figure 9.2. Disazo condensation pigments, developed by Ciba Geigy, are a range of durable yellow, red, violet and brown products, with structures such as compound **213** (C. I. Pigment Red 166). These pigments derive their name, and also their relatively high cost, from the rather elaborate synthetic procedures involved in their manufacture which involves a condensation reaction (see Scheme 3.8, Chapter 3). Hoechst has developed a series of azo pigments that contain the benzimidazolone group, *e.g.*

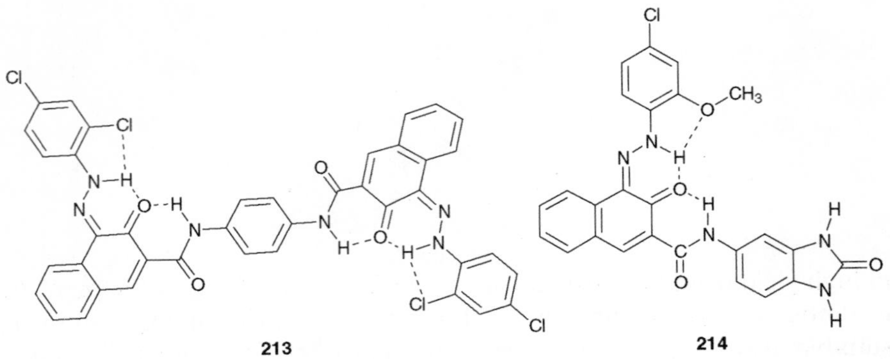

213 **214**

Figure 9.2 *Chemical structures of some high-performance azo pigments*

C. I. Pigment Red 183 (**214**), which range in shades from yellow to bluish-red and brown and exhibit excellent fastness properties. Their good stability to light and heat and their insolubility is attributed to extensive intermolecular association as a result of hydrogen bonding and dipolar forces in the crystal structure, as illustrated in Figure 9.3.

Carbonyl pigments of a variety of types may also be classed as high-performance products. These include some anthraquinones, quinacridones, perylenes, perinones, isoindolines and diketopyrrolopyrroles. The chemistry of these groups of colorant is discussed in Chapter 4. Some representative examples of chemical structures of important high-performance carbonyl pigments are illustrated in Figure 9.4.

A number of vat dyes developed originally for textile applications are suitable, after conversion into an appropriate pigmentary physical form, for use in many paint and plastics applications. Examples of these so-called vat pigments include the anthraquinones, Indanthrone Blue (**215**, C. I. Pigment Blue 60) and Flavanthrone Yellow (**216**, C. I. Pigment

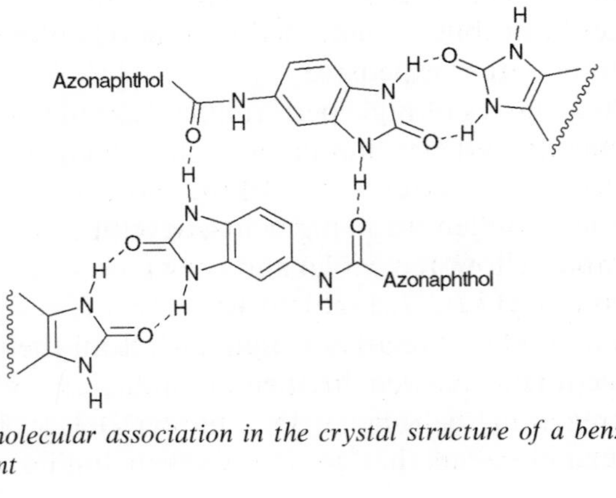

Figure 9.3 *Intermolecular association in the crystal structure of a benzimidazolone azo pigment*

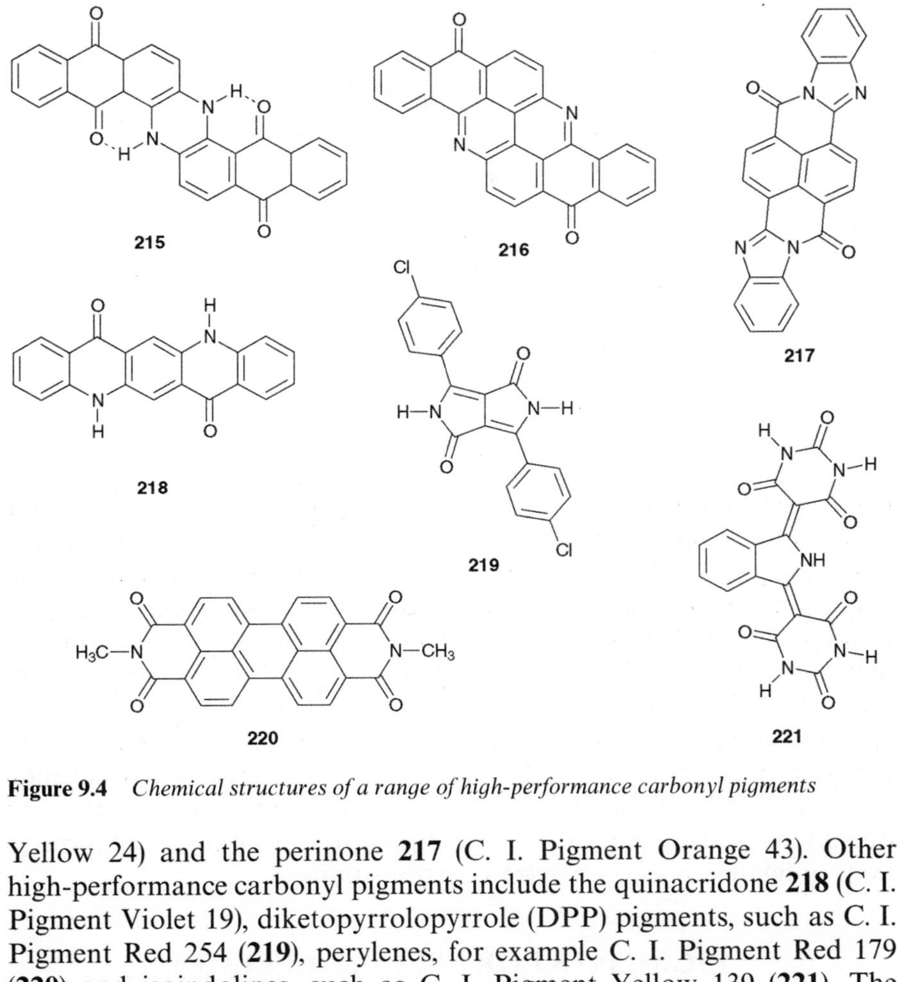

Figure 9.4 *Chemical structures of a range of high-performance carbonyl pigments*

Yellow 24) and the perinone **217** (C. I. Pigment Orange 43). Other high-performance carbonyl pigments include the quinacridone **218** (C. I. Pigment Violet 19), diketopyrrolopyrrole (DPP) pigments, such as C. I. Pigment Red 254 (**219**), perylenes, for example C. I. Pigment Red 179 (**220**) and isoindolines, such as C. I. Pigment Yellow 139 (**221**). The excellent lightfastness, solvent resistance and thermal stability of carbonyl pigments may be explained in many cases by intermolecular association in the solid state as a result of a combination of hydrogen bonding and dipolar forces, similar to that illustrated for the benzimidazolone azo pigments in Figure 9.4. A diagrammatic illustration of the intermolecular hydrogen bonding in the crystal lattice arrangement of quinacridone **218** is given in Figure 4.6, Chapter 4.

Other chemical types of high-performance organic pigments are exemplified by the tetrachloroisoindolinone **222** (C. I. Pigment Yellow 110) and the dioxazine **223** (Carbazole Violet, C. I. Pigment Violet 23). Considerable research has been carried out in an attempt to exploit the potential of metal complex chemistry to provide high-performance pig-

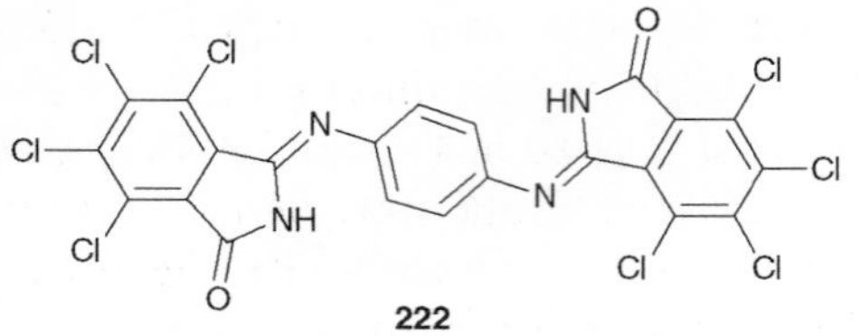

222

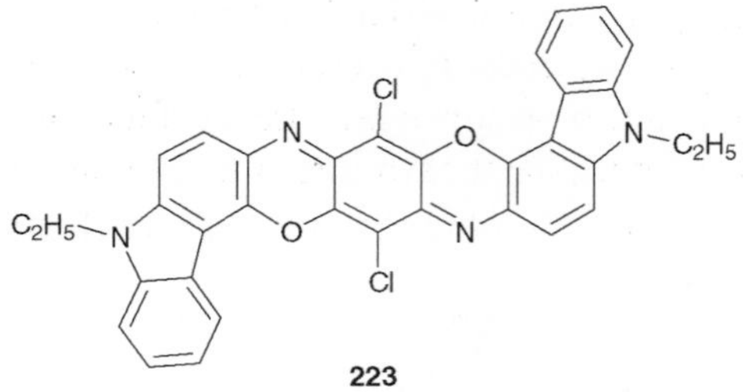

223

ments, particularly of yellow and red shades, to complement the colour range of the copper phthalocyanines (which cannot be extended outside blues and greens). A number of azo, azomethine and dioxime transition metal complex pigments have been obtained which show excellent light-fastness, solvent resistance and thermal stability. However, the products have achieved limited commercial success largely because the enhancement of the fastness properties of the organic ligand which results from complex formation is almost inevitably accompanied by a reduction in the brightness of the colour. This effect may be explained by a broadening of the absorption band as a result of overlap of the band due to π–π* transitions of the ligand with those due to transition metal d–d transitions or ligand–metal charge transfer transitions.

PIGMENTS FOR SPECIAL EFFECTS

Metallic, pearlescent and fluorescent pigments are grouped together in this section as three types of pigment used for their ability to produce unusual optical effects.

Metallic Pigments

By far the most important metallic pigment is aluminium flake, C. I. Pigment Metal 1. The use of aluminium pigments to provide the commonly-observed metallic effect in car finishes is well known, and they are used also in a range of printing ink and plastics applications. The pigments owe their importance to the highly reflective nature of aluminium metal and their stability, which is largely due to the thin, tenacious oxide film on the surface of the pigment particles. Aluminium pigments are generally manufactured from aluminium metal by a wet ballmilling process in the presence of a fatty acid (usually stearic or oleic) and a mineral oil. The presence of the liquid ingredients is essential to improve the efficiency of the process and to eliminate the potential explosion hazard of dry grinding. Aluminium pigments consist of small lamellar particles. The pigments are categorised according to their ability to 'leaf'. Leafing grades when incorporated into a film become oriented in a parallel overlapping fashion at or near the surface of the film, thus providing a continuous metallic sheath and a bright silvery finish. Non-leafing grades are distributed more evenly throughout the film producing a sparkling metallic effect.

Pearlescent Pigments

Pearlescent pigments give rise to a white pearl effect often accompanied by a coloured iridescence. The most important pearlescent pigments consist of thin platelets of mica coated with titanium dioxide which partly reflect and partly transmit incident light. Simultaneous reflection from many layers of oriented platelets creates the sense of depth which is characteristic of pearlescent lustre and, where the particles are of an appropriate thickness, colours are produced by interference phenomena. Pearlescent pigments are used in automotive finishes, plastics and cosmetics.

Fluorescent Pigments

Daylight fluorescent pigments consist essentially of fluorescent dyes dissolved in a transparent and colourless polymer. The resulting solid solutions are then ground to a fine particle size for incorporation as pigments into paints, printing inks or plastics. In these application media, the pigments in daylight give rise to colours which possess a remarkable vivid brilliance as a result of the extra glow of fluorescent light. The applications of daylight fluorescent pigments are generally associated with their extremely high visibility and their ability to attract attention such as in advertising and in the field of safety. Fluorescent pigments are mostly based on a toluenesulfonamide–melamine–formaldehyde resin matrix. The fluorescent dyes most commonly used in the pigments are rhodamines (Chapter 6) in the red to violet shade range, and aminonaphthalimides and coumarins (Chapter 4) in the yellow shade range.

Chapter 10

Functional or 'High Technology' Dyes and Pigments

The chemistry of the most important dyes and pigments used in the coloration of traditional substrates, including textiles, paints, printing inks and plastics, has been dealt with extensively in the previous chapters of this book. It is likely that this range of well-established products will remain for the foreseeable future as the most important materials manufactured for the purpose of providing colour. In recent decades, there is no doubt that new products for such conventional applications have appeared with reduced frequency as industry has concentrated research effort on process and product improvement, and addressing a range of environmental issues (Chapter 11). However, in the same period, there have been exciting developments in organic colour chemistry as a result of the opportunities presented by the emergence of a range of novel applications which place significantly different demands on dyes and pigments. These colorants have commonly been termed *functional*, because the applications in question often require the dyes or pigments to perform certain functions beyond the simple provision of colour. Alternatively, they may be referred to as *high technology* colorants, because they are designed for use in applications derived from advances in fields to which this particular term commonly refers. The applications of functional dyes and pigments include a wide range of electronic applications, including liquid crystal displays, microfilters, solar energy conversion, lasers and optical data storage, some of the more recently developed reprographic techniques, such as electrophotography and ink-jet printing, and a range of biomedical uses. While these colorants are unlikely to rival the traditional dyes and pigments in terms of the quantities required, they are potentially attractive to manufacturers due to the possibility of high added value. In this chapter, an overview of the principles of some of the more important of these applications is presented, necessarily selective in approach because of the diversity of the topics, together with a discussion in each selected case of the chemistry of the colorants which may be used. For some of these applications, traditional dyes and pigments may be used, although often the colorants may require special purification procedures, not a common requirement for conventional applications. For others, new colorants tailored to the needs of the particular application have been designed and synthesised.

ELECTRONIC APPLICATIONS OF DYES AND PIGMENTS

The rapid advances in electronics technology which were a feature of the latter part of the 20th century have had an immense impact on our lives. The numerous examples of these developments include the growth in the ownership and in the sophistication of personal microcomputers, the wide range of electronically-controlled goods which are now commonplace in the home and in offices, including kitchen appliances, televisions, video recorders and hi-fi equipment, digital cameras and numerous electronic components in cars. The list is virtually endless and there is little evidence to suggest that rate of development of new electronic products will slow down in years to come. It may be a cliché, but we are indeed living in the electronic age. Examples of electronic applications in which dyes and pigments play an important functional role include liquid crystal displays, microcolour filters, lasers, solar energy conversion, electrochromic displays and optical data storage.

Dyes for Liquid Crystal Displays

Liquid crystals, commonly referred to as the fourth state of matter, are materials which are intermediate in character between the solid and liquid states. Unlike normal isotropic liquids, they show some time-averaged positional orientation of the molecules, but they retain many of the properties of liquids, such as the ability to flow. In recent decades, liquid crystals have played an increasingly important part in our lives. Probably their most familiar application is in the information displays which provide the visual interface with microprocessor-controlled instrumentation. Liquid crystal displays have superseded more traditional display technology, such as light-emitting diodes and cathode ray tubes, for many appliances principally because of the advantages of visual appeal, low power consumption, and their ability to facilitate the miniaturisation of devices into which they are incorporated. They are encountered widely in many applications such as digital watches, calculators and instrument display panels in cars and aircraft.

Liquid crystal displays operate by utilising the ability to control ambient light to provide contrast, *i.e.* areas of dark and light, within the display. This is achieved as a result of a change in orientation of the liquid crystal molecules within certain sections of the display as an electric field is applied. In early devices a polariser was used to provide this contrast, but it was discovered that improved contrast could be provided by the use of certain specifically-designed dyes which are dissolved in the liquid crystal host material. These *dichroic* dyes are believed to align with the liquid crystal molecules and change direction with the application of the electric field, as illustrated in Figure 10.1, with a consequent change in colour intensity.

To explain the change in colour intensity with change in orientation of the dye with respect to the direction of incident light indicated by Figure 10.1, it is important to recognise that electronic excitation caused by light absorption is accompanied by a change in the polarity of the dye molecule. The electronic transition which occurs may be ascribed a transition dipole moment which has not only magnitude but also direction. The transition moment in aminoazobenzene dye **224**, for example, is directed from the electron-releasing amino group to the electron-withdrawing nitro group, as illustrated Figure 10.2 (see Chapter 2 for a justification based on the application of the valence-bond approach to colour and constitution). It is of particular note that, in the case of dye **224**, the transition moment is more or less aligned with the long axis of the molecule. For light absorption to occur, the electric vector of the incident light (perpendicular to the direction of propagation of light as illustrated

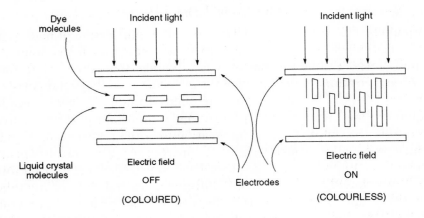

Figure 10.1 *Orientation of dyes in on/off states of a liquid crystal display*

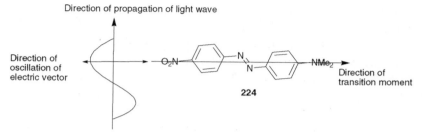

Figure 10.2 *Orientation of an aminoazobenzene dye for maximum light absorption in a liquid crystal display*

in Figure 10.2) must oscillate in the same direction as the transition moment of the dye. If the transition moment of the dye is perpendicular to the direction of the light source, *i.e.* parallel to the electric vector, then maximum colour intensity will be achieved. If the molecule rotates through 90° when the field is switched, either on or off, then a change to minimum colour intensity will be observed.

For application in liquid crystal displays, the dyes require good solubility in and compatibility with the liquid crystal host, and thus the dyes are invariably non-ionic. The dyes are also required to be of high purity and good lightfastness. Most important of all from the point of view of providing high contrast is the ability of the dye molecule to align, and switch, with the liquid crystal host. This ability may be quantified in terms of the *order parameter*, *S*, a measurable quantity ($= 0$ for non-alignment and $= 1$ for perfect alignment). The target value for a good liquid crystal display dye is $S \cong 0.8$. Generally, liquid crystal displays are required to produce black/white contrast, and thus to ensure that absorption occurs throughout the visible spectrum, a combination of yellow, red and blue dyes is required. Azo dyes, such as compound **224** are capable of providing only reasonable order parameters and also generally suffer from inadequate light stability. The dyes which most effectively satisfy the requirements for liquid crystal displays belong to the carbonyl chemical class. Anthraquinone dyes, such as the yellow dye **225**, the red **226** and the blue **227** (illustrated in Figure 10.3) are especially suitable for these applications. It is commonly perceived that a rod-like molecular shape provides the required orientation behaviour for liquid crystal dyes. This feature is apparent in the structure of blue dye **227** and, additionally, the extensive intramolecular hydrogen bonding in this dye promotes good lightfastness. However, this type of molecular shape is not so apparent in the sulfur-containing anthraquinone dyes **225** and **226**. These examples demonstrate that the nature of the interactions between the dye and the liquid crystal molecules in these displays is probably more complex than

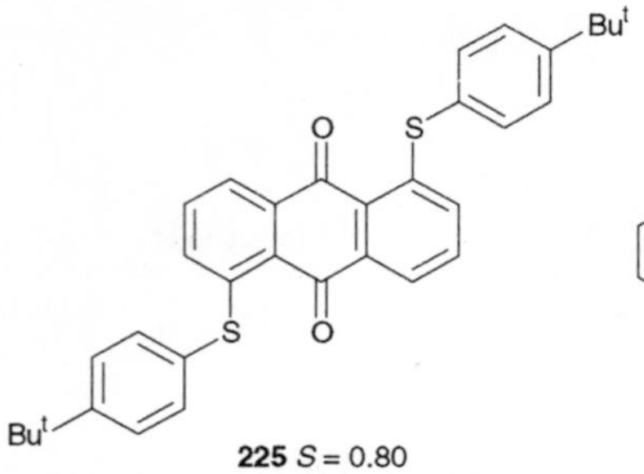

Figure 10.3 _Anthraquinone dyes used in liquid crystal displays_

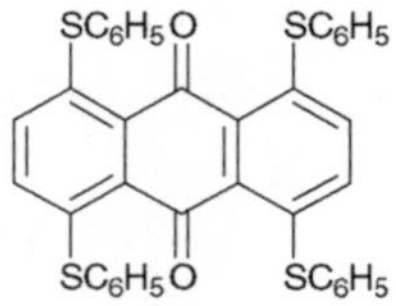

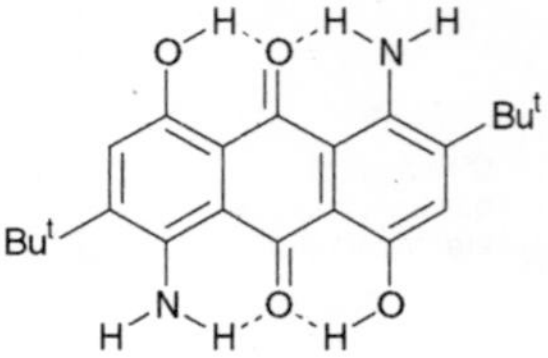

that depicted in this simplistic approach and as yet must be considered as incompletely understood.

Pigments for Microcolour Filters

Flat-screen displays, such as those used in miniature televisions and computer displays normally produce a multicolour effect by means of microcolour filters used together with thin film transistors and liquid crystals, commonly also with fluorescent back-lighting. Since the observer is directly viewing the source of the light, the process involves the use of additive colour mixing and hence requires filters which incorporate the additive primary colours – red, green and blue (Chapter 2). Pigments are generally preferred to dyes for these applications because of their superior durability. The microfilters are manufactured by a process in which pigments of the appropriate three colours are deposited by vacuum sublimation on to a matrix of an appropriate design produced by photo-microlithography. The pigments for colour filter applications require the ability to sublime, good thermal stability and lightfastness, and appropriate spectral characteristics. Suitable pigments include the perylene **228** (red), copper phthalocyanine **229** (blue) and its octaphenyl derivative **230** (green) (Figure 10.4).

Laser Dyes

Lasers are devices which provide an intense, continuous source of 'in-phase' radiation. The term _laser_ is an acronym referring to _light_ amplification by _stimulated_ emission of radiation. Traditional laser technology utilises a variety of inorganic materials to produce the required emission. Several different types of inorganic laser have been developed to emit either in the ultraviolet region, the visible region or infrared region of the

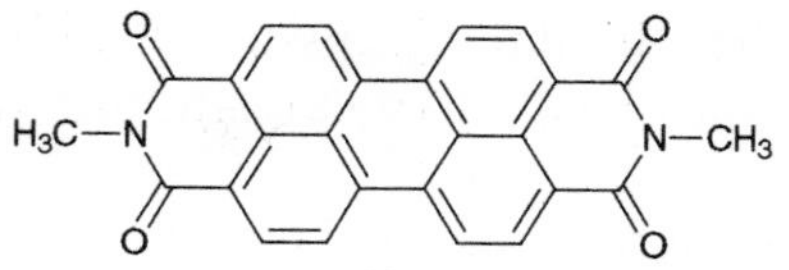

Figure 10.4 *Organic pigments used in colour filters for flat-screen displays*

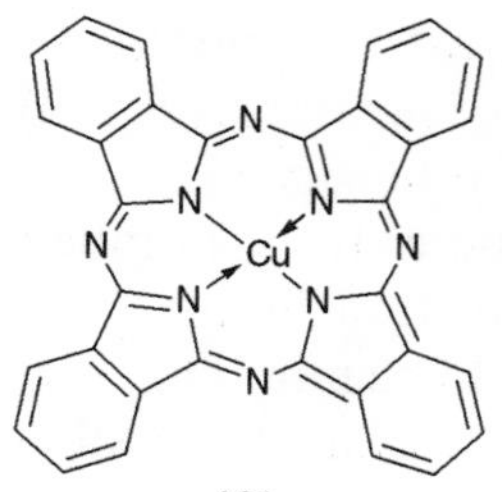

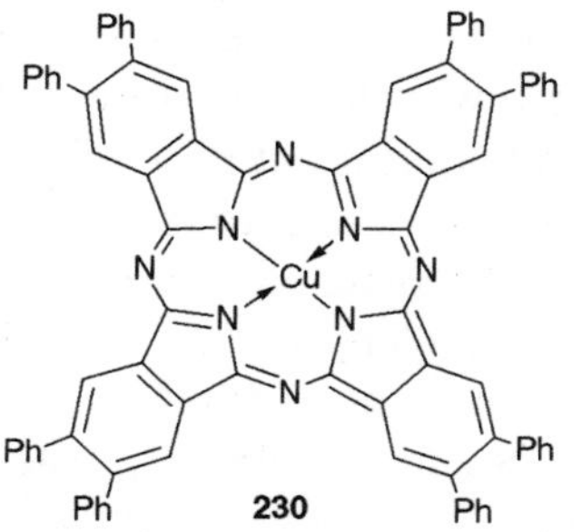

electromagnetic spectrum. These inorganic lasers are commonly low cost and robust devices, but they suffer from the disadvantage that they emit at only a few selected wavelengths, and in very narrow bands. An example of a commonly encountered inorganic laser is the gallium–arsenic diode laser, which emits in the near-infrared region at around 780 nm. In contrast, dye lasers, which are based on specific fluorescent dyes dissolved in an appropriate solvent, emit a broad band of wavelengths, thus offering the advantage of tunability through a wide range of wavelengths. The applications of dye lasers include communications technology, microsurgery, spectroscopy, photochemistry, studies of reaction kinetics, isotope separation and microanalysis.

The principles of the function of the fluorescent dye in a dye laser may be explained with reference to the mechanism of fluorescence which is described in Chapter 2 (see Figure 2.6). Having absorbed a quantum of light, the dye molecule is promoted from its ground state, S_0, to its first excited state, S_1^*. Lasing occurs when incident radiation interacts with the dye molecule in its excited state, thereby causing (stimulating) the molecule to decay to the ground state by emission of radiation (fluorescence). For the lasing effect to occur, a situation has to be brought about in which the dye molecules exist predominantly in the excited state. This population inversion is achieved by 'pumping' the system with an appropriate source of energy, generally a powerful inorganic laser. In contrast to spontaneous emission, the stimulated emission of radiation from dye lasers is strictly coherent (same phase and polarisation) and is of high intensity.

There are a number of general requirements for laser dyes. Strong absorption at the excitation wavelength is clearly required but there should be minimal absorption at the lasing wavelength, so that there is as little overlap as possible between absorption and emission spectra. Additional significant requirements are a high quantum yield (0.5–1.0), a short fluorescence lifetime (5–10 ns), low absorption in the first excited state at the pumping and lasing wavelengths, low probability of intersystem crossing to the triplet state and good photochemical stability. By appropriate dye selection it is possible to produce coherent light of any wavelength from 320 to 1200 nm. Examples of laser dyes are given in Figure 10.5. For shorter wavelengths (up to *ca*. 470 nm), aromatic hydrocarbons, such as anthracenes and polyphenyls, and fluorescent brightening agent-type materials such as stilbenes, oxazoles, coumarins and carbostyrils are most commonly used. Benzimidazolylcoumarins such as compound **231** are suitable for use in the 470–550 nm range, while rhodamines, such as compound **232**, together with some related xanthene derivatives, are of prime importance in the 510–700 nm region of the visible spectrum. For dye lasers operating at longer wavelengths into the near-infrared region, oxazines and polymethine dyes are most important. For example, polymethine dye **233** provides a lasing maximum at 950 nm.

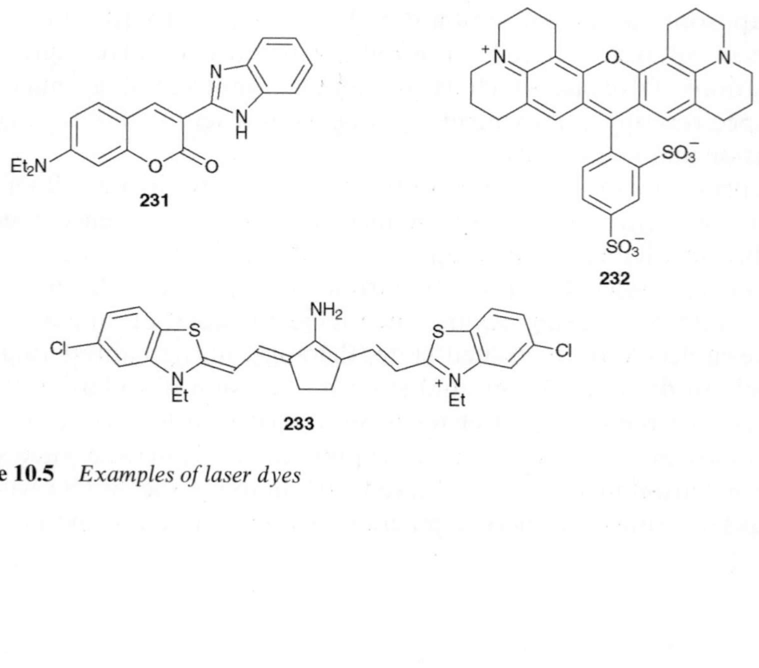

Figure 10.5 *Examples of laser dyes*

Dyes in Solar Energy Conversion

There has been considerable effort over many years aimed at the development of the means to convert solar energy into electrical energy. The potential advantages of solar power are obvious. It utilises a non-diminishing energy source and suffers little from the global environmental problems inherent in energy generation by nuclear fission or by the combustion of fossil fuels.

Dye-sensitised photocells, which utilise specific dyes for the purpose of solar energy conversion, have been investigated as an alternative to traditional silicon photocell technology. One type of construction for such a photocell is shown in Figure 10.6. The electrodes for the photocell are commonly thin films of metal (*e.g.* aluminium or gold) coated on to glass. At one electrode, there is a semiconducting interfacial layer of an inorganic substrate (*e.g.* fine particle-size or 'nano-structured' titanium dioxide), on to which the dye is adsorbed. Another important component of the photocell is an electrolyte, dissolved in an appropriate organic solvent and attracted into the intra-electrode space by capillary action. The electrolyte is commonly an iodine/tetraalkylammonium iodide mixture.

A mechanism which has been proposed for the operation of this type of photocell is illustrated in Figure 10.7, although it is not fully established in detail. In the proposed mechanism, it is suggested that absorption of light by the dye (or sensitiser, S) raises the dye to its first excited state S*. In the excited state, S* releases an electron into the conducting band of the titanium dioxide electrode, at the same time forming oxidised sensitiser, S^+. At the counter-electrode, an electron is transferred to the

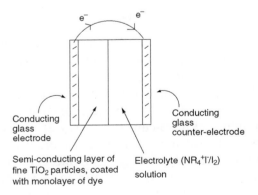

Figure 10.6 *Design of a dye-sensitised photocell*

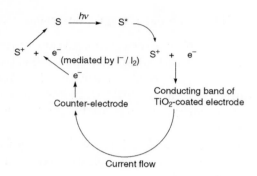

Figure 10.7   *Proposed mechanism for the operation of a dye-sensitised photocell*

electrolyte and, mediated by the iodine/iodide equilibrium, the dye (S) is regenerated by reduction.

A dye which shows particular promise for this application is the octahedral ruthenium(II) complex of 2,2′-bipyridyl (**234**). While this type of system appears to offer considerable potential as a means of solar energy conversion, the efficiency of the technology, at its current state of development, is significantly lower than that of traditional silicon photo-cells.

234

One of the deficiencies of conventional silicon cells used for solar energy conversion is their low efficiency at lower wavelengths in the UV and visible region. An approach to the solution of this problem has involved the development of dye-based solar collectors. These solar collectors utilise the ability of fluorescent dyes to absorb lower wavelengths of light and to re-emit the energy at higher wavelengths to which the photocells are more sensitive. In practice, solar collectors contain the fluorescent dye in a thin sheet of plastic, generally poly(methyl methacrylate), whose faces and edges are mirrored to chan-nel the emitted radiation by internal reflection to the edge of the sheet containing the solar cell as illustrated in Figure 10.8. The requirements of

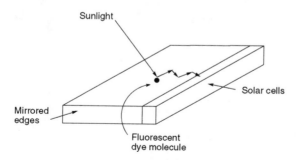

Figure 10.8   *Design of a fluorescent dye-based solar collector*

dyes for this application are a high fluorescence quantum yield, compatibility with the plastic material, high lightfastness in the polymer matrix, little overlap of the absorption and emission spectra (generally associated with a large Stokes shift) and high purity. A wide range of dye classes have been examined for their suitability for application in solar collectors and the search for suitable dyes, especially for dyes with appropriate photostability, is continuing. Coumarins and perylenes are currently the most suitable classes of dye for these applications.

Near-infrared Absorbing Dyes

In recent years there has been growing commercial interest in compounds which absorb in the near-infrared region of the electromagnetic spectrum (*i.e.* beyond 750 nm), a region which formerly had been regarded mainly as a scientific curiosity, rather than offering practical application significance. Interest in this region emerged particularly as a result of the development of relatively inexpensive and robust inorganic semiconductor lasers, such as the gallium–arsenic diode laser which emits around 780–830 nm, and the potential which these devices offered for the development of optoelectronic devices. Compounds which absorb at these wavelengths are commonly referred to as near-infrared absorbing dyes, even though in the strict sense they are not dyes since their absorption is outside the visible region. Nevertheless, these compounds are commonly designed and synthesised by an extension of conventional dye chemistry so that it is convenient to consider them in this context.

Optical data storage methods offer a number of advantages over conventional information storage methods, including high storage capacity and environmental friendliness. Methods have been developed which utilise the ability of near-infrared emitting inorganic lasers to write and

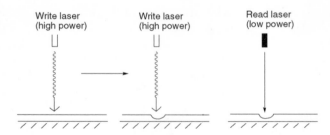

Figure 10.9   *Principles of optical data storage using near-infrared absorbers*

read information. The principles of one such method, referred to as the WORM (*w*rite *o*nce *r*ead *m*any) system, are illustrated in Figure 10.9. The optical recording media consist of discs on which a thin film of a polymer containing a near-infrared absorbing dye is deposited. When the radiation from a laser of sufficient power impacts the disc, the dye absorbs the radiation. Conversion of the absorbed radiation to heat generates a temperature which is sufficient to melt the polymer and cause the formation of a pit. The pattern formed by the pits constitutes the data. To read the information contained in the pits, the disc is scanned with a laser of lower power, insufficient to cause melting. Instead, a detector records the amount of radiation reflected from the disc, the reflectivity being lower in the data pits. The first generation compounds for optical data storage were inorganic materials, such as tellurium alloys, which were expensive and highly toxic. As a wide range of near-infrared absorbing organic dyes became available, improved performance in such applications was possible. Subsequent to these developments, the CD-R system was developed which offered the advantage over the WORM disks of compatibility with standard compact disk technology. The CD-R system uses dyes with a $\lambda_{\max}$ in the range 690–710 nm and with a small but finite absorbance in the range 780–830 nm. Optical data storage development has subsequently focused on the DVD-R system, which exploits the introduction of low cost solid state lasers operating in the 630–650 nm range and which thus utilises dyes absorbing at lower visible light wavelengths. Other applications which have emerged for near-infrared absorbing dyes include security printing, for example in banknotes and 'invisible' but laser-readable bar-coding for the identification of branded products, infrared photography, laser filters and appropriately designed compounds have potential for use in photodynamic therapy, a cancer treatment which is described in a later section of this chapter. Examples of near-infrared absorbing dyes include a range of phthalocyanine derivatives such as the polyarylthiophthalocyanines **235**, croconium dyes such as compound **236** and nickel dithiolene complexes such as compound **237**.

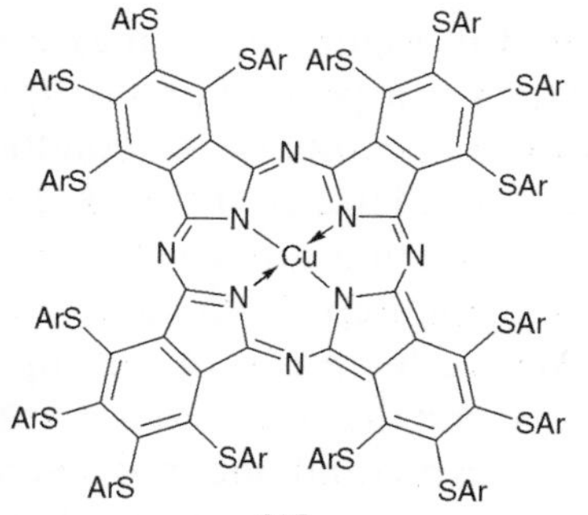

235

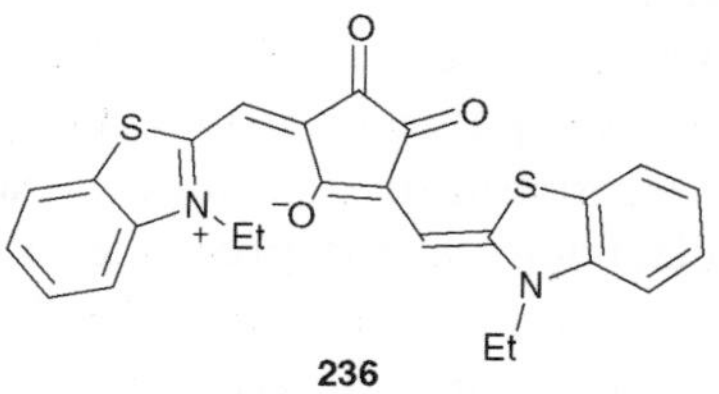

236

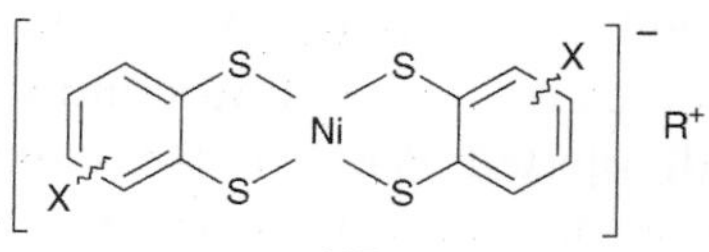

237

REPROGRAPHICS APPLICATIONS OF DYES AND PIGMENTS

The term *reprography* was first used around the 1960s to encompass the new imaging techniques, including xerography, electrofax and thermography, which were emerging at the time for document reproduction. Reprographics still refers to those newer imaging processes which may be distinguished from conventional printing techniques, such as lithography, flexography and gravure printing, and traditional silver halide photography. These processes have experienced remarkable growth because of the rapid advances in computing technology and the consequent growth in demand for home and office printing of high quality, using relatively inexpensive printers. The most important of these techniques currently are electrophotography and ink-jet printing, which offer the advantage of printing on to plain paper. Another technology worthy of mention in this context is dye diffusion thermal transfer (D2T2), but the need for special paper and the consequent expense has limited the growth of this technique.

Electrophotography

The term *electrophotography* encompasses the familiar techniques of photocopying and laser printing. In these printing systems, the ink is a toner, which is generally a powder consisting mainly of pigment, charge control agent, and a low melting binder. Toner printing systems use optical or electrical methods to form an electrostatic latent image to which the toner is attracted and subsequently transferred to the substrate.

Most photocopiers and laser printers operate on a similar basis, so that it is convenient to discuss them together, but there are also some key differences.

The photocopying process, as illustrated in Figure 10.10, may essentially be separated into six stages. One of the key aspects of photocopying is a photoconducting material which, as the name implies, is a conductor of electricity in the presence of light but an insulator in the dark. In the first stage of the process, the dark photoconducting surface of a drum is given a uniform electrical charge. In the imaging step, the document to be copied is illuminated with white light. Where there is no print, light is reflected on to the photoconducting surface and this causes the charge to be dissipated. Where there is text, light is absorbed and does not reach the photoconductor. In this way, a latent electrostatic image is formed on the drum. The drum is then exposed to toner particles, which have been given an opposite charge to that on the drum. The toner particles are thus attracted to the image areas of the photoconducting surface. The next step involves transfer of the image from the drum to the substrate, usually plain paper, by applying to the back of the paper an electrical potential of opposite charge to that of the toner particles. The next step fixes the image to the paper. This is achieved by a heat treatment, which melts the toner resin and fuses it to the paper. Finally, the photoconducting surface is cleaned to make it ready for the next copying cycle.

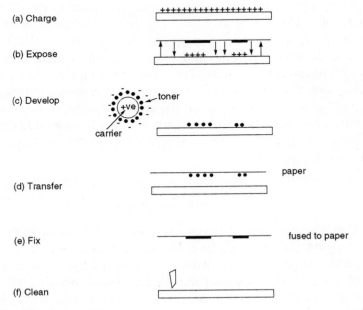

Figure 10.10   *The photocopying process*

In a laser printer, the first step also involves giving the photoconducting drum a uniform electric charge. However, in the second step a laser is used to write on the charged drum the information stored in the memory of the printer. In contrast to the photocopying process, therefore, the charge is dissipated in the image areas. In this case, the toner used carries a charge which is the same as that on the drum, so that it is repelled from the non-image areas on to the uncharged image areas. From this stage onwards, the principles of the laser printing process are similar to the photocopying process.

Organic colorants of a variety of types are key components in the operation of photocopiers and laser printers. The photoconducting drum contains two surface layers. The first is a thin charge-generating layer which is on top of a thicker charge-transport layer. The charge-generating layer contains a charge-generating material which is usually a highly-purified pigment, commonly of the perylene, polycyclic anthraquinone or phthalocyanine type, with carefully controlled crystallinity and particle size. Light interacts with the pigment to form an ion-pair complex at the same time releasing an electron which passes to earth. A positive hole is thus generated which migrates to the interface with the charge-transport layer. This layer contains highly electron-rich compounds, such as poly-arylamines, not usually coloured materials, which facilitate transfer of the positive holes to the outer surface. The toners used in photocopying and laser printers are powders which consist mainly of pigments, charge control agents and resin, a relatively low melting (60–70 °C) polymer. The most commonly used pigment is carbon black since most printing remains monochrome. However, the production of multicoloured prints by this method using pigments of the three subtractive primary colours, yellow, magenta and cyan is increasing in popularity. The yellows are commonly provided by disazo pigments, the magenta by quinacridones and the cyan by copper phthalocyanine (see Chapter 9 for a description of the chemical structures of these types of pigments). Charge control agents are materials which, as the name implies, assist in the control of the electrostatic charge applied during the printing process. These are ionic materials (either anionic or cationic) which may be coloured. For example, 2:1 chromium complexes of azo dyes, similar to those used as anionic premetallised dyes for application to wool (see Chapter 7) are commonly used in black toners.

Ink-jet printing

Ink-jet printing is currently experiencing tremendous growth. It is a non-impact means of generating images which involves directing small droplets of ink in rapid succession, under computer control, onto the substrate. There are a number of types of ink-jet printing methods. Two of the principal types are the continuous jet method and the impulse or 'drop on demand' method, although only the latter is considered here. In this method (Figure 10.11) pressure on the ink is applied to form a droplet when it is needed to form part of the image. An array of nozzles is used to generate the image and the print-head is required to be as close as possible to the substrate surface so as to produce an accurate image.

Ink-jet printing requires the use of inks which meet stringent physical, chemical and environmental criteria. Ink-jet printing inks are very low viscosity fluids as is required by the non-impact method of ink delivery. They have a remarkably simple composition consisting of a solvent, almost invariably water, and a colorant together with other additives for specific purposes. In contrast to most other printing inks, the colorants used are mainly dyes because pigments, even when an extremely fine particle size is used, have a tendency to block the nozzles. Initially, the dyes used in ink-jet printing were water-soluble dyes selected from the range of conventional dyes used in textile applications, notably from the acid, direct, or reactive dye application classes, or from food dyes, this last group offering the particular advantage of clearly-established non-toxicity. An example of such a 'first generation' ink-jet dye is C. I. Food Black 2. (**238**) (Figure 10.12). These early ink-jet dyes performed reasonably in many respects, but the principal problem was inadequate water-fastness which could, for example, lead to smudging of the print when handled with moist fingers. In the 'second generation' ink-jet dyes, this particular feature was improved, by the design of specific dyes which are soluble at the slightly alkaline pH (7.5–10) of the ink, but which are rendered insoluble by the weakly acidic pH conditions (pH 4.5–6.5) on the paper

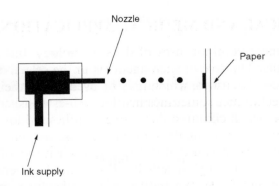

Figure 10.11   *'Drop-on-demand' ink-jet printing*

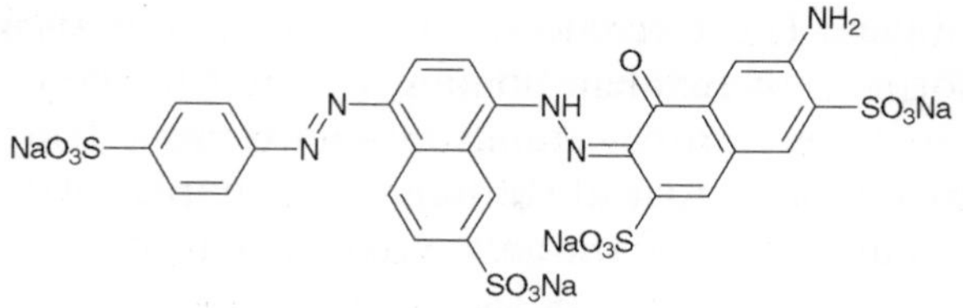

Figure 10.12 *Examples of water-soluble dyes used in ink-jet printing*

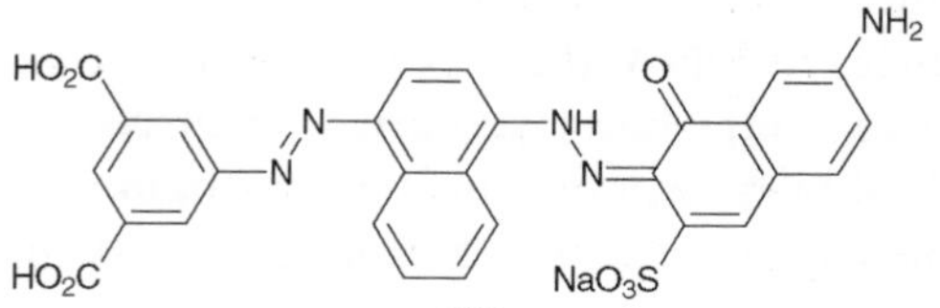

substrate. One of the simplest ways in which this was achieved involved the incorporation of carboxylic acid groups, which are ionised (as $-CO_2^-$) at the higher pH, thus enhancing water solubility, but are protonated (as $-CO_2H$) as the pH is reduced, thus reducing solubility. An example of such a dye is compound **239** (Figure 10.12), whose molecular structure is closely related structurally to C. I. Food Black 2 (**238**) but with sulfonic acid groups replaced, to a certain extent, with carboxylic acid groups. Dyes of this type are capable of providing prints which are reasonably resistant to wet treatments. Ink-jet printing is currently used extensively for home and office printing, where high-quality text and graphics reproduction is required. The technology also offers considerable potential to make significant inroads in the future into higher volume industrial printing, for example for packaging, textiles, wall-coverings and advertising displays.

BIOLOGICAL AND MEDICAL APPLICATIONS OF DYES

One of the most common uses of dyes in biology and medicine is as staining agents. The role of such agents is in the selective coloration of certain biological features, which may be used to aid their characterisation. These techniques sometimes involve well-established dyes, but frequently make use of coloured derivatives synthesised for a specific purpose. The most important dyes in current use as staining agents are strongly fluorescent because of the ease and sensitivity of detection and their ability to provide high specificity. The methods generally involve the selective binding of the dye molecules to substances, tissue, cells or micro-organisms and subsequent detection by fluorimetric or microscopic techniques. The mechanism of the binding may involve either physical adsorption or covalent bonding but in many cases it is not fully established. Fluorescent dyes are also used extensively in the detection and identification of biological substances. Examples of such analytical applications, which most commonly have a diagnostic purpose, include the detection of specific antibodies, the identification of nucleic acids, proteins, lipids and polysaccharides in cells, the identification of specific nucleotide sequences in DNA, the study of cell membranes and the early detection of cancer cells. The dyes are thus useful as probes for the investigation of biological activity and the mechanism of biological reactions. Research using such probes, particularly in the study of biomembrane properties, has already led to the development of new diagnostic procedures for certain diseases. In addition to selectivity of binding, the fluorescent dyes used must be non-toxic, exhibit a high quantum yield and a specific emission spectrum which is sufficiently distinct from background fluorescence. Since fluorescence in the blue–violet region is often inherent in biological systems, dyes which fluoresce in the orange, red and green regions are most usefully employed. Among the most widely used fluorochromes in biomedical applications are fluorescein, rhodamines, acridines and their derivatives.

Photodynamic Therapy

Arguably the most significant recent medical development which involves the use of dyes is photodynamic therapy (PDT). PDT is a treatment for cancer that uses a combination of laser light, a photosensitising compound (the dye), and molecular oxygen. In this treatment, the photosensitiser is given intravenously to the patient and some time is allowed (3–96 h) so that it equilibrates within the body. During this time the photosensitiser penetrates into the tumour cells. Irradiation of the cells with laser light may then initiate their destruction, thus providing a potential means for destroying the tumour. The method is rapidly gaining acceptance as a treatment for certain forms of cancer, which shows significantly fewer side-effects than conventional treatments. The photophysical mechanism of PDT is illustrated in Figure 10.13. Laser light interacts with the photosensitiser and promotes it from its ground state (^{1}Sens) into its singlet excited state (^{1}Sens*), which then undergoes intersystem crossing to a triplet excited state (^{3}Sens*). This in turn interacts with molecular oxygen which, because its ground state is a triplet, returns the sensitiser to its ground state and in the process generates singlet oxygen (^{1}O$_2$). Singlet oxygen is the effective photodynamic agent which is highly reactive towards, for example, unsaturated centres in the proteins

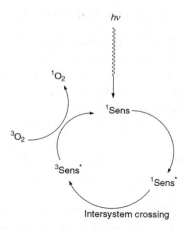

Figure 10.13 *Photophysical mechanisms involved in photodynamic therapy*

and lipids which construct the cell membrane. The most commonly used photosensitiser for PDT is referred to as Photofrin. To produce Photofrin, haematoporphyrin, obtained from blood, is first treated with acetic and sulfuric acids followed by an alkaline work-up to give a derivative, HPD, which is a complex mixture of porphyrin monomers and oligomers. Photofrin is obtained from HPD by removal of the non-active monomeric components. There has been a vast amount of research carried out in the search for improved photosensitising agents, and the work is ongoing. Among the range of requirements for such an agent are that it should ideally absorb at $>600\,nm$ for improved absorption of the laser light, improved specificity for tumour cells, and a triplet excited state which is sufficiently long lived to facilitate the formation of singlet oxygen. The dye should be soluble in body fluids and should clear from the body. A variety of modified porphyrins have been investigated for this purpose and particular promise has also been shown by a number of phthalocyanine derivatives, for example the sulfonated aluminium(III) phthalocyanine **240**.

240

CHEMICHROMISM

Chemichromism is a term which has been introduced to describe a change in colour, caused by a chemical conversion, induced by some external stimulus, for example exposure to light, heat or electric current. Examples of these phenomena have been well known for many years but, for many traditional applications such as dyes applied to textiles, colour changes, for example when exposed to light, were regarded as a nuisance and highly undesirable. As time has progressed, and as potential niche applications have been recognised, there has been a resurgence of interest in dyes which exhibit phenomena of this type, especially when the effects are reversible and controllable.

Probably the most familar and longest-established use of reversible colour change chemistry is found in titration indicators. The best known of these compounds are the indicators used traditionally to detect the end-point in an acid–base titration, as a result of a colour change, which takes place within a specific narrow pH range. As an example, the azo dye, methyl orange (**241**) changes from orange to red in the pH range 3–4, due to conversion into the protonated species (**242**) (Scheme 10.1). Other types of sensor which make use of specific colour changes include redox indicators, which show a reversible colour change as a result of electron transfer reactions, and chromoionophores, which give rise to specific colour changes when they react with certain metal ions and thus may be used as sensitive reagents for metal ion detection.

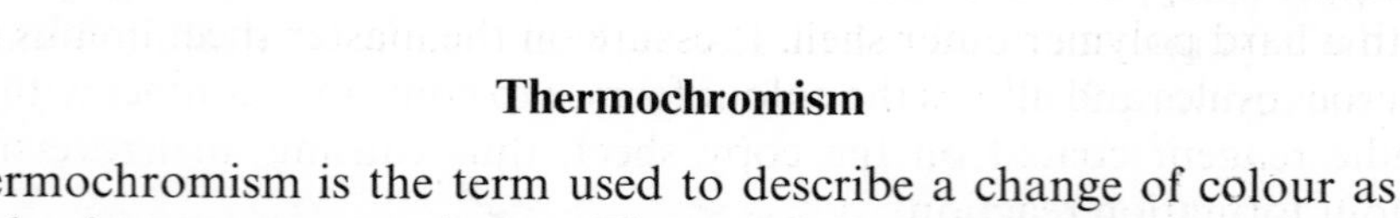

Scheme 10.1   *Colour change in the protonation of methyl orange (**241**), an acid–base indicator*

Thermochromism

Thermochromism is the term used to describe a change of colour as a result of a temperature change. Thermochromic dyes and pigments find application where this colour change is used to indicate a temperature change, for example in plastic strip thermometers, medical thermography and non-destructive testing of engineered articles and electronic circuitry. They may also be used in thermal imaging and for a variety of decorative or novelty effects. Crystal violet lactone (**243**) is one of the best-known components of thermochromic colorants, and may be used as an example to illustrate the principles. Compound **243** is a non-planar molecule which is colourless. In contact with acid, the lactone ring is opened to give the violet/blue arylcarbonium ion (triarylmethine) structure (**244**), as illustrated in Scheme 10.2. Similar chemistry is exhibited by certain xanthene derivatives (fluorans), represented by structure **245** in Scheme 10.2, and these compounds enable an extension of the available colour range in the ring-opened forms (**246**) to orange, red, green and black.

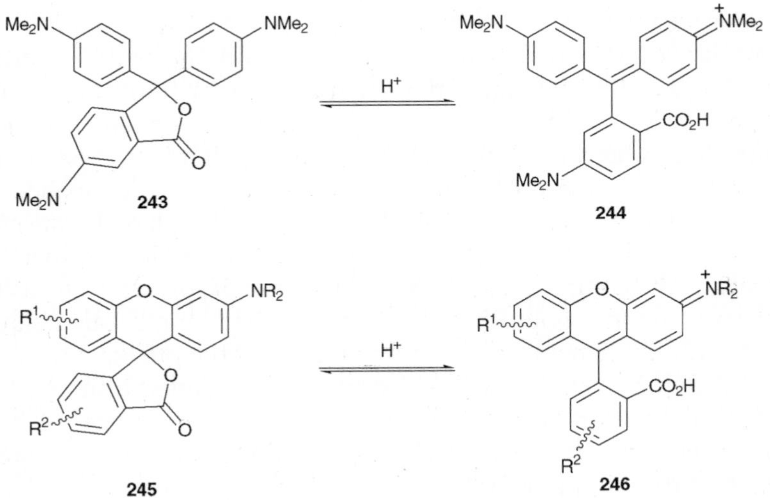

Scheme 10.2 *Colour formation in the ring-opening of crystal violet lactone and related compounds*

Colour formation reactions of this type are utilised in carbonless copy paper, which is based on the principle of colour formation on the copy as a result of pressure of writing or typing in the master sheet. In such systems, the underside of the master sheet contains the colour former, for example compound **243**, encased in microcapsules, which are tiny spheres with a hard polymer outer shell. Pressure on the master sheet breaks the microcapsules and allows the colour former to come into contact with an acidic reagent coated on the copy sheet, thus causing an irreversible colour formation reaction.

Colour formers such as compounds **243** and **245** are not inherently thermochromic. For example, they melt without any change in colour. However, they may be used to generate colour thermally, either irreversibly or reversibly, as composite materials. In thermally sensitive paper, the colour former and an acidic developer, usually a phenol, are dispersed as insoluble particles in a layer of film-forming material. When brought into contact with a thermal head at around 80–120 °C, the composite mixture melts as a result of the localised heating. As a result, the colour former and developer diffuse together and react to form a colour. This process is assisted by the presence of third component, a sensitiser, such as dibenzyl terephthalate, which assists diffusion by acting as a solvent. In this case, the process is irreversible and a permanent image is formed.

Reversible thermochromic systems of this type are also known in which a composite with a similar set of the three ingredients, colour former, developer and solvent (in this case a relatively non-polar solvent such as a fatty acid), are contained together in microcapsules. In these systems, it is curious that, even though essentially the same chemistry is involved, the composite material is usually coloured at low temperatures and decolourises as the temperature is raised. To explain this effect, a mechanism has been tentatively proposed in which it is suggested that at low temperatures the coloured protonated cationic ring-opened form exists as an ion-pair complex with the anion of the phenolic developer. This complex is insoluble in the low polarity solvent at low temperatures. On heating, the complex dissolves, and because of solvation effects associated with the low polarity of the solvent, the complex dissociates to form the colourless neutral ring-closed form and the phenol. On cooling, recrystallisation occurs and this results in the regeneration of the coloured ion-pair complex.

Thermochromism is also shown by certain liquid crystal materials. Chiral nematic liquid crystals adopt a helical structure and, when the pitch length of the helix is of the same order of magnitude as the wavelength of light, colours may be produced from the incident white light by an interference effect. The exact wavelength of the reflected light is dependent on the pitch length. Thermochromism is observed because of the variation in helical pitch length with temperature. The visual effect commonly involves a continuous change through a spectrum of colours as the temperature is raised, and is most striking when viewed against a black background which absorbs the transmitted wavelengths.

Photochromism

Photochromism is generally regarded as the reversible interchange of a single compound between two molecular states, which show different absorption spectra when activated by light. Most commonly, reversible photochromism follows the general principles illustrated in Scheme 10.3 in which structure A (usually colourless) is converted by irradiation with light of a certain wavelength (usually in the UV region) into structure B which is coloured. The reverse reaction takes place either thermally when the light source is removed or by irradiation with light of a different

Structure A $\xrightleftharpoons[\Delta \text{ or } h\nu_2]{h\nu_1}$ Structure B

(colourless)   (coloured)

Scheme 10.3   *General scheme for reversible photochromism*

wavelength. Colour changes due to exposure to light have been known for many years. The term *phototropy* was formerly used to refer to the change of colour of dyes on textile fabrics when exposed to light, which has always been considered as undesirable even if it is reversible. This effect was particularly evident in some simple aminoazobenzene derivatives used as the first generation of disperse dyes applied to cellulose acetate. The effect was attributed to the photo-induced conversion of the *trans* isomer of the azo dye into its *cis* isomer and the associated colour change.

The modern generation of photochromic dyes are used in sun-screening applications, such as sunglasses and car sunroofs, optical data storage, photoresponsive polymers and for a variety of novelty effects. One of the most important groups of photochromic dyes are the spirooxazines. Spirooxazine **247** is non-planar and colourless. On exposure to light, compound **247** ring opens to the violet–blue merocyanine structure **248** (Scheme 10.4). When the light source is removed, it reverts thermally to the more stable ring-closed structure **247**. The spirooxazines offer the advantage of good fatigue resistance and relative ease of synthesis.

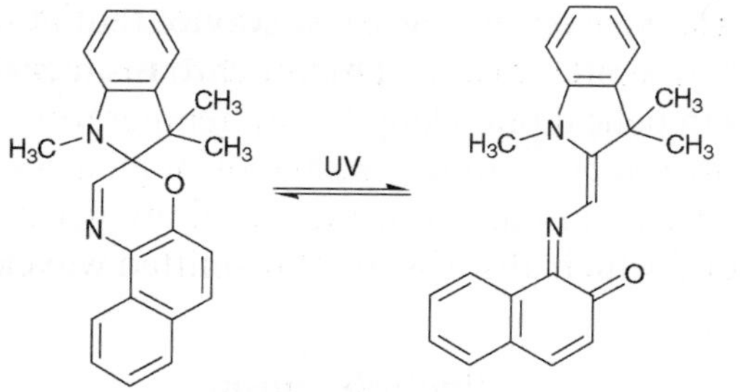

Scheme 10.4   *Reversible photochromism of spirooxazine* **247** *involving interconversion with the merocyanine form* **248**

Electrochromism

Electrochromic dyes, as the name implies, are dyes which undergo a colour change as a result of the application of electrical energy. These dyes are of obvious interest because of their potential for application in displays. One approach to the production of electrochromic displays involves the chemistry illustrated in Scheme 10.5. This approach makes use of colourless biscationic materials, such as the paraquat derivatives **249**. As illustrated in the scheme, these compounds are reduced electrochemically to a coloured radical cation **250**, which is deposited at an electrode. One of the principal difficulties which remains to be overcome commercially for displays using this chemistry is the ageing process which causes the deposited material, eventually, to crystallise, thus inhibiting the reverse oxidation process.

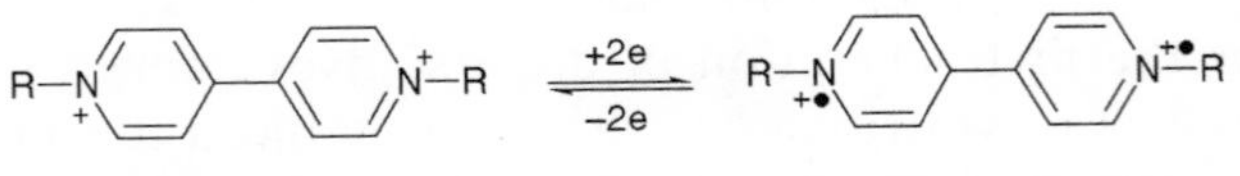

249 colourless   **250** coloured

Scheme 10.5   *The reversible electrochromism of paraquat derivatives*

An alternative approach to the production of electroluminescent displays has emerged from the discovery that poly(*p*-phenylenevinylene) (PPV), **251**, produces a greenish glow when a thin film of the polymer is subjected to a high voltage. Since this discovery, a large number of light emitting polymers, mostly based on highly conjugated materials of the PPV type, have been prepared and investigated in the search for polymers with improved electroluminescence efficiency and which emit at a range of different wavelengths. There is no doubt that this research paves the way for significant future developments in flat-screen display technology.

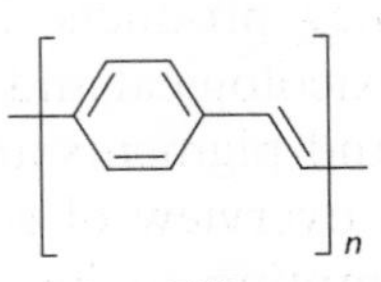

251

Chapter 11

Colour and the Environment

Over the last few decades, society has become increasingly sensitive towards the protection of the environment. We have developed a concern over a range of issues, amongst the most extensively debated being the destruction of the rain forests, global warming and the depletion of the ozone layer. Over this period, the chemical manufacturing industry has been faced with the need to address its responsibility towards a wide range of health, safety and environmental issues, and indeed, probably its most significant current challenges are associated with the requirement to satisfy the demands of stringent environmental controls. Concern about the potential adverse effects of the chemical industry on the environment is global nowadays, although the response in some parts of the world has been much faster and more intense than in others. The colour manufacturing industry represents a relatively small part of the overall chemical industry. It may be described, in general, as a small-volume, multi-product industry and it has traditionally been highly innovative, constantly seeking to introduce new products. This has led to a requirement to address a wide range of toxicological and ecotoxicological issues both in the manufacture of dyes and pigments and in their application. This chapter seeks to present an overview of some of the more important general issues, without attempting to be comprehensive, because the diversity of chemical types and applications of dyes and pigments means that the range of individual issues is immense. A detailed discussion of the legislation and controls introduced by governments and regulatory agencies, the introduction of which has been a major factor in ensuring compliance of the industry with the most important issues, is outside the scope of this chapter, because of the complexity of the legislative detail, the fact that it varies substantially from country to country and because the situation is constantly evolving, so that information presented would quickly become out of date.

191

The colour manufacturer and user has a duty to address environmental and toxicological risks from a variety of points of view, including hazards in the workplace, exposure of the general public to the materials and the general effect on the environment. The level of risk from exposure to chemicals is clearly of prime concern for those handling materials in large quantities in the workplace, and this has required the introduction of modern work practices to minimise the exposure. The approach towards addressing the problems associated with exposure to potentially danger-ous chemical substances, which has been adopted in most countries, commonly involves an evaluation of risk, including an assessment of the hazards presented by the various chemical species on the basis of the available toxicological data, and an assessment of exposure levels, and, from this evaluation, risk management strategies are developed.

Dyes and pigments are, by definition, highly visible materials. Thus, even minor releases into the environment may cause the appearance of colour, for example in open waters, which attracts the critical attention of the public and local authorities. There is thus a requirement on industry to minimise environmental release of colour, even in cases where a small but visible release might be considered as toxicologically rather innocu-ous. A major source of release of colour into the environment is asso-ciated with the incomplete exhaustion of dyes onto textile fibres from an aqueous dyeing process and the need to reduce the amount of residual dye in textile effluent has thus become a major concern in recent years. While this applies in principle to all application classes of textile dyes, the particular case of reactive dyes for wool and cellulose is of special interest because of the problem of dye hydrolysis, which competes with the dye–fibre reaction, the unfixed and hydrolysed dye inevitably appearing in the effluent. An extensive programme of research to address these problems has met with some success, leading to the development of more selective fibre-reactive systems and significantly improved processing conditions (Chapter 8). However, the development of a practical reactive dye system which is completely free of the problems associated with hydrolysis or incomplete dye–fibre reaction has so far proved elusive. An alternative approach to addressing the problem of colour in textile dyeing effluent has involved the development of effluent treatment methods to remove colour. These methods inevitably add to the cost of the overall process and some present the complication associated with the possible toxicity of degradation products. A number of chemical treatment methods for colour removal have been developed, of which the most successful involve oxidative degradation, for example using chlorine or ozone. Ozone treatment is particularly effective but is rather expensive. Physical treatments may also be used to remove colour, for example by the adsorption of dyes onto inert substrates such as activated charcoal, silica, cellulose derivatives or ion-exchange resins. Biological processes make use of the ability of living organisms to bind or degrade colour. Biodegradation processes offer the attraction of the potential for the decoloration of effluent with complete mineralisation of the organic materials present to carbon dioxide, water and inorganic ions such as nitrate, sulfate and chloride. However, the principal problem with the development of such a system is that synthetic dyes are generally xenobiotic, *i.e.* are not metabolised by the enzymes in the micro-organisms present naturally in waters, and the solution may require the development of micro-organisms cultured specifically for the purpose of metabolising synthetic dyes. The problems associated with textile effluent are not restricted to the dyes themselves. There are also issues associated with the presence of traces of heavy metals, notably the excess chromium used in the chrome mordanting of wool, the high levels of inorganic salts and other auxiliaries required by certain textile dyeing processes, and the use of reducing agents in the case of sulfur dyes. Each specific process presents its own set of environmental issues which are required to be addressed by the development of new products, process improvements or effluent treatment methods (Chapter 7). For pigments, because of their insolubility and the particular way they are used, the loss into the environment is much less than with textile dyes.

The importance of being aware of the potential adverse effects of exposure to chemicals on our health is self-evident. Toxic effects may be categorised in a number of ways. Acute toxicity refers to the effects of short-term exposure to a substance, for example in a single oral administration. It is relatively reassuring that studies of textile dyes suggest that there is little evidence for acute oral toxicity, and that most show little or no toxicological effect. Only in the cases of a few cationic dyes and some disazo dyes have some significant toxic effects been suggested. Pigments and vat dyes generally show remarkably low acute toxicity. This is generally attributed to their extremely low solubility in body fluids, which means that they are capable of passing through the digestive system without absorption into the bloodstream.

Chronic toxicity refers to the effect of regular exposure over a prolonged period of time. Arguably the most severe chronic toxicological effect is the potential to induce cancer. There has been considerable concern in recent years over the potential carcinogenicity of certain azo dyes, which centres on the possibility of metabolism of the dyes by reductive cleavage of the azo group to give the aromatic amines from which they are derived synthetically. A number of aromatic amines which have been used in colour manufacture are recognised carcinogens. In particular, it is well established that benzidine (**252a**) and 2–naphthylamine (**253**) are potent human carcinogens. Epidemiological studies carried out in the first half of the 20th century demonstrated a pronounced increase in the incidence of bladder cancer in workmen employed in the dye manufacturing industry who had been exposed to these two amines. As a consequence, responsible manufacturers in the Western world discontinued the manufacture of dyes from these amines, although the manufacture continued in certain other parts of the world. It has emerged subsequently that benzidine-derived azo dyes, such as Congo Red, C. I. Direct Red 1 (**254**), may also present a carcinogenic risk, an effect attributed to the metabolism of the dyes to benzidine by enzymatic reduction. Certain European countries have examined the cancer-causing potential of azo dyes critically, by focusing on the amines which would be released if reductive cleavage of the azo group were to take place. In Germany, for example, the approach has been to ban the manufacture and importation of all dyes derived from a list of twenty aromatic amines believed to be animal carcinogens. On the other hand, it is by no means true that all azo dyes should be considered as potential carcinogens. Studies of structure–carcinogenicity relationships in azo dyes have demonstrated, for example, that when the amine produced by reductive cleavage of the azo group contains a sulfonic acid group there is little or no carcinogenic risk. Azo pigments are considered to present a considerably reduced risk compared with azo dyes, so that, for example, they are not included in the German legislation. 3,3′-Dichlorobenzidine (DCB), **252b**, is on the German list of carcinogenic amines, yet it continues to be used in the manufacture of a series of important commercial disazo pigments (Diarylide Yellows and Oranges, see Chapter 9), since there does not appear to be strong experimental evidence that these pigments are metabolised to DCB. However, there is evidence that the

252a X = H
252b X = Cl

253

254

Diarylide pigments may cleave thermally with prolonged heating above 240 °C to give a series of decomposition products which include DCB. Since this evidence emerged, there has been a decline in the use of these pigments in applications where high temperatures are likely to be encountered, for example in thermoplastics. Exposure to some types of dyes can give rise to rather less severe chronic toxicity effects, for example contact dermatitis and, especially for certain reactive dyes, respiratory sensitisation.

There are a number of interesting issues associated with the colours present in foods, and their influence on our health. Most foods contain natural colouring materials and, in addition, it is common to add colour to enhance the appeal of some foods to the consumer. There is growing evidence that certain natural colorants, for example the carotenoids present in fresh fruit and vegetables (Chapter 6), are of therapeutic value as a consequence of their anti-oxidant properties, which, it is suggested, may provide protection against cancer. There is an increasing tendency to use natural dyes as food additives, but synthetic dyes are more commonly used as they offer the advantage of higher colour strength and lower cost. A relatively restricted range of synthetic dyes are considered acceptable for use as food colorants, mainly because of the stringent toxicological requirements. An important example of a synthetic food colorant is the monazo dye Tartrazine, C. I. Food Yellow 23 (255). It is of note that reductive enzymatic cleavage of the azo group in this dye would lead to sulfonated aromatic amines, which are considered to present little toxicological risk. An interesting development in food colours is a range of polymeric dyes, such as the anthraquinone dye 256. It is considered that these dyes present little risk because their high molecular weight means that they do not penetrate through the membranes of the gastrointestinal tract.

255

256

The extension of the use of natural dyes into an even wider range of applications, including textiles and leather, as alternatives to synthetic dyes would seem superficially to offer attractive possibilities. In the current debate on the issues involved, the range of opinions expressed is extremely wide. There are environmental arguments which support the use of natural dyes, as this is seen to be exploiting renewable resources thus making a contribution towards environmental sustainability, at the same time presenting minimal risk to human health. There are, however, a number of counter-arguments. Natural dyes are generally more expensive and show inferior technical performance compared with synthetic dyes. In addition, the large-scale production and use of natural dyes might well introduce environmental problems, for example the need to cultivate large areas of land and the likelihood that the coloration processes would not necessarily be free from pollution problems. A possible outcome might be a renewed interest in natural dye research towards improved product performance, production methods, and application processes, and the emergence of certain niche markets which take advantage of the positive environmental perception associated with natural colours.

The issue of the so-called 'heavy metals' is of some concern within the chemical industry. While the term 'heavy metal' might have originally been derived on the basis of density, this property alone does not necessarily relate directly to toxicity or environmental behaviour. In fact, certain metals, for example iron, zinc, manganese, copper, chromium, molybdenum and cobalt, might technically be described as 'heavy', yet they are essential for life. The term has come to be used to represent those metals which are regarded as detrimental to the environment when a certain concentration is exceeded. The elements mercury, cadmium and lead occupy a special position because the concentration at which they begin to become detrimental to the health of organisms is very low. In addition, when absorbed in excessive amounts they may accumulate and can cause chronic health problems. Mercury is not a constituent of any significant commercial colorant. However, one important use of mercury derivatives is as a catalyst in the sulfonation of anthraquinone, an essential first step in the synthesis of some anthraquinone dyes (Chapter 4). Special care is required to ensure that mercury is not released into the environment in the effluent from such processes. Inorganic pigments based on lead (lead chromates and molybdates) and cadmium (cadmium sulfides and sulfoselenides) are still used commercially (Chapter 9). Their use nowadays is restricted significantly by a series of voluntary codes of practice, reinforced by legislation in certain cases, for example in toy finishes, graphic instruments and food contact applications where ingestion is a possibility. Cadmium pigments are used mainly in the coloration of certain engineering plastic materials which require very high temperature processing conditions. At present, completely satisfactory substitutes for such applications are not available, especially in terms of thermal and chemical stability. Lead chromates continue to be used, mainly in coatings, because they remain by far the most cost-effective, high durability yellow and orange pigments. Furthermore it may be argued that lead and cadmium pigments do not present a major health hazard or environmental risk, because of their extreme insolubility. Nevertheless, it seems likely that the progressive replacement of these products by more acceptable inorganic and organic pigments will continue in an increasing range of applications. Chromium(VI) is also considered to be highly toxic. In addition to its presence in lead chromate pigments, dichromates are used in the dyeing of wool by the chrome mordanting process (Chapter 7). This process remains of some importance as a cost-effective means of producing deep colours, which are fast to light and washing, on wool. The continuing acceptability of this dyeing method owes much to recent process improvements, for example the development of dyeing auxiliaries and procedures which minimise the level of residual excess chromium in textile dye effluent. There are toxicological and environmental issues associated with a number of other metals used to a certain extent in colorants, for example barium, manganese, cobalt, nickel, copper and zinc, but at present their use remains acceptable.

Certain organic chlorine derivatives, most notably the polychlorobiphenyls (PCBs), are known to be highly toxic and give rise to considerable environmental concerns. The potential to form PCBs in trace amounts has been noted in the manufacture of certain colorants, for example when aromatic chloro compounds are used either as reactants or solvents. In such situations, the industry has been required to respond by developing processes either to eliminate PCB formation or to ensure that the levels are below the rigorous limits set by legislation. Some environmental agencies and activist groups have taken a more extreme view and advocated a complete ban on all chlorinated organic chemicals. Such a ban would have a major effect on the colorant industry because of the large number of organic dyes and pigments which contain chlorine substituents. So far, legislation to this effect has not been introduced. Indeed, any move in such a direction would be an over-reaction, since there is little evidence that the mere presence of chlorine in a molecule means that it poses an environmental risk.

Environmental arguments are highly emotive. There are many reasons why the public has become sensitive to environmental issues and, to a certain extent, suspicious of the attitudes of the chemical industry in this respect. The industry has made errors in the past, some with severe human and environmental consequences, and has had to examine its conscience on occasions for having paid inadequate regard to health, safety and environmental issues. However, there is a point of view that the pendulum may have swung to the opposite extreme, so that, for example, arguments based on perception, rather than solid scientific evidence, often win the day. The need for rigorous experimentation to address toxicological and environmental concerns has acted as a deterrent to innovation in terms of the introduction of new products, in view of the considerable expense required in testing before a product may be introduced into the marketplace. We should not lose sight of the fact that the main reason for the existence of the synthetic colour industry is that dyes and pigments enhance our environment, by bringing attractive colours into our lives. It would indeed be a dull world if, for example, television, movies, photography and magazines essentially only provided black and white images, and if automobiles were available in 'any colour we like provided it's black!' The days when such a situation existed are not so long ago in relative historical terms. Equally, it is vital that those industries involved in the manufacture and application of colour should continue to be sensitive to any potential adverse effect on the environment in its wider sense, and respond accordingly. A balanced approach will ensure protection of the environment and allow an innovative colorant industry to thrive in the 21st century.

Bibliography

E. N. Abrahart, *Dyes and their Intermediates*, Edward Arnold, London, 1977.

R. L. M. Allen, *Colour Chemistry*, Nelson, London, 1971.

The Chemistry of Synthetic Dyes, ed. K. Venkataraman, Academic Press, London, 1952–1978, vols. 1–8.

R. M. Christie, *Pigments: Structures and Synthetic Procedures*, Oil and Colour Chemists Association, London, 1993.

R. M. Christie, R. R. Mather and R. H. Wardman, *The Chemistry of Colour Application*, Blackwell Scientific, London, 1999.

J. Fabian and H. Hartmann, *Light Absorption of Organic Colorants: Theoretical Treatment and Empirical Rules*, Springer-Verlag, New York, 1980

P. F. Gordon and P. Gregory, *Organic Chemistry in Colour*, Springer-Verlag, New York, 1983.

P. Gregory, *High-technology Applications of Organic Colorants*, Plenum Press, London, 1991.

The Chemistry and Application of Dyes, ed. D. R. Waring and G. Hallas, Plenum Press, London, 1990.

Colorants and Auxiliaries, ed. J. Shore, Society of Dyers and Colourists, Bradford, 1990 vol. 1.

J. Griffiths, *Colour and Constitution of Organic Molecules*, Academic Press, London, 1976.

J. Griffiths (ed.), 'Developments in the chemistry and technology of organic dyes', *Critical Reports on Applied Chemistry 7*, Blackwell Scientific Publications, London, 1984.

W. Herbst and K. Hunger, *Industrial Organic Pigments: Production, Properties and Applications*, 2nd edn., VCH, Weinheim, 1997.

W. Ingamells, *Colour for Textiles. A User's Handbook*, The Society of Dyers and Colourists, Bradford, 1993.

R. McDonald, *Colour Physics for Industry*, 2nd edn., The Society of Dyers and Colourists, Bradford, 1997.

Environmental Chemistry of Dyes and Pigments, ed. A. Reife, John Wiley, Chichester, 1996.

K. McLaren, *The Colour Science of Dyes and Pigments*, Adam Hilger, Bristol, 1986.

A. H. M. Renfrew, *Reactive dyes for Textile Fibres: the Chemistry of Activated π-bonds as Reactive Groups and Miscellaneous Topics*, Society of Dyers and Colourists, Bradford, 1999.

Industrial Inorganic Pigments, 2nd edn., ed. G. Buxbaum, Wiley-VCH, Heidelberg, Berlin, 1998.

Colour for Science, Art and Technology, ed. K. Nassau, Elsevier, Amsterdam, 1998.

P. Rys and H. Zollinger, *Fundamentals of the Chemistry and Application of Dyes*, John Wiley, London, 1972.

J. D. Saunders, *Pigments for Inkmakers*, SITA Technology, London, 1989.

The Colour Index, 3rd revision, Society of Dyers and Colourists, Bradford, 1988, vol. 1–9.

Pigment Handbook, vol. 1: Properties and Economics, ed. P. A. Lewis, John Wiley, Chichester, 1988.

H. Zollinger, *Color: a Multidisciplinary Approach*, Wiley-VCH, Heidelberg, Berlin, 1998.
Color Chemistry: Synthesis, Properties and Applications of Organic Dyes and Pigments, 2nd
    edn., VCH, Weinheim, New York, 1991.

Subject Index